Introduction to Modern Colloid Science

D1382930

Introduction to Modern
Colloid Science

ROBERT J. HUNTER

OXFORD NEW YORK MELBOURNE
OXFORD UNIVERSITY PRESS

Oxford University Press, Walton Street, Oxford OX2 6DP

Oxford   New York
Athens   Auckland   Bangkok   Bombay
Calcutta   Cape Town   Dar es Salaam   Delhi
Florence   Hong Kong   Istanbul   Karachi
Kuala Lumpur   Madras   Madrid   Melbourne
Mexico City   Nairobi   Paris   Singapore
Taipei   Tokyo   Toronto

and associated companies in
Berlin   Ibadan

Oxford is a trade mark of Oxford University Press

Published in the United States
by Oxford University Press Inc., New York

© Robert J. Hunter, 1993
First published 1993
Reprinted with corrections 1994, 1996

A catalogue record for this book is available from the British Library

Library of Congress Cataloging in Publication Data
Hunter, Robert J.
Introduction to modern colloid science / Robert J. Hunter. — 1st ed.
Includes bibliographical references.
1. Colloids. I. Title.
QD549.H945   1993   541.3'45 — dc20   93–12783
ISBN 0 19 855386 2 (Pbk)

Printed in Great Britain by
Bookcraft (Bath) Ltd, Midsomer Norton

PREFACE

About seven years ago, a few of my colleagues and I decided to write a textbook of colloid science which would assume that the reader had no prior knowledge of the subject but had a solid grasp of physical chemistry. The idea was to provide such a person with an introduction to the research literature. The result was the two volumes of *Foundations of colloid science*, published by Oxford University Press.

The underlying approach was to try to make all the theoretical work accessible to a researcher prepared to make the necessary effort. Having seen the end result, I felt that it should be possible to convey a lot of the more important material from that work at a less demanding level, suitable for a senior undergraduate course, or for those many workers in science and industry for whom colloid science is important, but not central, to their concerns.

The result is the present work. It draws very heavily on some parts of *Foundations*, but there is material here which is not in the earlier work. It uses the same technique of posing many problems designed to fill out the algebraic derivations of formulae, but there are also problems of the more traditional kind. The treatment is rather more mathematical than has been attempted before at this level, but the mathematical methods are introduced carefully and nothing more than the most elementary calculus is required to understand the text. Although there are numerous references to applications throughout the text, there is a final chapter devoted solely to that subject, in order to bring some of the concepts together and to illustrate the wide utility of colloid and surface science.

To keep the book within bounds there has, perforce, been much selection necessary. For those who wish to pursue some areas further, there is ample reference to the larger text and the terminology and symbolism have been kept consistent with *Foundations* as far as possible.

I have tried to get some of the concepts over by simple line drawings. These I have drawn myself with the help of the Drawperfect™ program. I hope that the crudeness of my artistry does not confuse the message too much.

The modern word processor has made it possible for me to prepare the typescript myself but the final production owes much to the patience of the people at Oxford University Press.

In the March 1996 reprint the answers to selected exercises have been included. A complete set of solutions is available from the author by contacting him by e-mail on hunter_r@chem.usyd.edu.au.

Sydney   R.J.H.
November 1992

CONTENTS

1

CHARACTERIZATION OF COLLOIDAL
DISPERSIONS

1.1 Nature of the colloidal state

The subject of colloid science covers a wide range of seemingly very different systems, from shaving cream to paints and cosmetics, from beer to agricultural soils, and from mayonnaise to biological cells. What do these systems have in common? They all consist of small 'particles' of one substance distributed more or less uniformly throughout another. This latter phase is continuous whilst the 'particles' are discontinuous. The continuous phase may be a gas, a liquid, or a solid whilst the discontinuous phase can also be a gas, liquid, or solid. The system will be colloidal if the 'particles' are sufficiently small and that usually means less than about 1 μm in at least one important dimension.

There are some important natural and synthetic systems in which solid or liquid particles are dispersed in a gas (these are called aerosols) and a few in which the continuous phase is solid (as in some mineral ores and alloys) but there are many more important systems in which the dispersion (continuous) phase is a liquid, and it is on these latter systems that we will concentrate attention, except for a few special situations.

When one substance dissolves in another to form a true solution the ultimate particles of solute are of molecular dimensions. The diameter of the solute molecules is in these cases at most a few nanometres, and more usually a few hundred picometres, which is comparable with the size of the solvent molecules. In colloidal systems, by contrast, the particles which correspond to the solute are very much larger than those of the solvent or continuous phase. Such systems may arise in the following ways:

1. The size of a solute molecule may be very much larger than the size of the solvent molecules. This is the case, for example, with solutions of proteins or polysaccharides (like starch) in water. Indeed the name 'colloid' means gum-like and was given to these substances by Thomas Graham in the early nineteenth century because the polysaccharide gums were among the first substances of this type which he subjected to systematic study.

2. A large number of solute molecules may associate together to form an aggregate which is much larger than the individual solvent molecules. If the bonds between the solute molecules are normal (covalent) chemical bonds, we regard the resulting large molecule as a polymer and this system is then the same as (1) above. If, however, the solute molecules associate by much weaker physical interactions (such as van der Waals forces, Chapter 9), the aggregate has very special properties and we refer to this as an 'association colloid'. The solutions formed by soaps and detergents fall into this category; they are sometimes referred to as colloidal electrolytes. We will examine some of their properties in §1.4.2.

3. A substance which is quite insoluble in a particular solvent may be broken down into very small particles which can then be dispersed throughout the 'solvent'. A finely divided precipitate of silver chloride crystals or a clay mineral suspension in water would be of this type. Such a system is regarded as colloidal if the size of the particles falls anywhere in the range 1–1000 nm. The reason for this rather arbitrary definition is that for smaller particles the systems show little or no difference from ordinary (true) solutions. The upper limit is set by the fact that almost all of the special features of colloidal systems may be traced to their having a very large surface area in contact with the dispersion medium. For particles larger than about 1000 nm (that is, 1 μm) the surface or **interfacial region** becomes of rather less significance compared to the bulk properties.

Table 1.1 The various types of colloidal dispersion with some common examples

Disperse phase	Dispersion medium	Technical name	Common name
Solid	Gas	Aerosol	Smoke, dust
Liquid	Gas	Aerosol	Fog
Solid	Liquid	Sol or colloidal sol	Suspension, slurry, jelly
Liquid	Liquid	Emulsion	Emulsion
Gas	Liquid	Foam	Foam, froth
Solid	Solid	Solid dispersion	Some alloys and glasses
Liquid	Solid	Solid emulsion	
Gas	Solid	Solid foam	

It is not necessary that all of the dimensions of the dispersed particles be very small for the system to be of interest to colloid scientists. Some important colloidal systems have particles which can readily be seen with the aid of an ordinary microscope, such as textile fibres or the cellulose fibres which make up filter paper or writing paper. It is sufficient if one of the characteristic dimensions of the particle (in this case the fibre diameter) falls within the stated range. In the case of a foam, such as that on the 'head' of a beer, it is not the bubbles of gas but the thin film of liquid between them which has the relevant dimension. Indeed, the only real requirement is that the surface area be large and if one admits the possibility of porous solids which can have an 'internal' surface it is possible for particles which are visible to the naked eye to be of interest to colloid science. This is the case with the molecular sieves and some solid catalysts, but in most such cases one would be able to break down the visible particle to submicroscopic size without affecting its properties very drastically.

Table 1.1 gives a summary of the various possible types of dispersion. Only the gas/gas dispersion is missing since no attempt has yet been made to study such systems. They could have at most a transient existence under rather high pressure conditions since rapid interdiffusion would presumably destroy them.

Exercises

1.1.1 Give a domestic example of each of the first six classes of dispersion listed above and state its approximate composition (for example, milk is an emulsion of fat (oil) in water). Bitumen (macadam) would be an example of a solid emulsion. Ice-cream is both a solid emulsion and a solid foam. The solid foams used as insulating materials in refrigeration, in cool boxes, and for packaging hot take-away food are examples of gas/solid dispersions.

1.1.2 Starting with a cube of solid, 1 cm along each edge, what is the total surface area when the solid is subdivided into cubes 1 μm along each edge? Repeat the calculation for a 0.1 μm cube and a 0.01 μm cube. Calculate the surface energy per particle in each case, assuming that the surface energy is 70 mJ m^{-2} and compare this to thermal energy (kT) at room temperature (25°C). What is the total surface energy for each system?

1.1.3 Kaolin (a clay mineral) particles are disc shaped with a diameter to thickness ratio of about 10:1. Estimate the surface area per gram of such a system when the average particle diameter is 1.2 μm assuming the density is 2.8 g cm^{-3}. If an aqueous suspension contains 85 g of solid per litre, what is the volume fraction of solids?

1.1.4 Derive an expression for the surface area per gram of a disc-shaped colloid, in terms of the solid density, ρ, the disc radius, a, and the aspect (diameter to thickness) ratio, R.

1.2 Biological and technological significance of colloids

Apart from the obviously colloidal nature of protein and polysaccharide solutions there are many other biological systems which have been studied by the methods of colloid science. The flow properties of blood are best understood in terms of its being a colloidal dispersion of (deformable) flat plates (the red corpuscles) in a liquid. The flow properties of material in the lower digestive tract must sometimes be modified by colloid chemical techniques to treat constipation or diarrhoea. The synovial fluids which lubricate the joints and bearing surfaces in the body owe their remarkable properties to their colloidal character. Also the adhesion between cells and the interaction between antigens and antibodies is currently being treated by the same mathematical theory that applies to the coagulation of colloidal particles. The greater part of the food processing, preserving, and packaging industry rests heavily on colloid chemistry, and agricultural scientists require a knowledge of the colloidal properties of soils in order to induce optimum plant growth.

Almost all of the ancient and modern craft industries also draw most of their technical expertise from colloid science. In paper-making, both the cellulose fibre used as the meshwork, and the clay used as a filler to improve opacity and produce a shiny texture, are colloidal. The inks used in ball-point pens and also printer's ink owe their special flow properties to their colloidal constituents as do also the many varieties of paint and cosmetics (Chapter 10). Ceramic products from expensive china to building bricks are made from clay/water sols and most of the plastics and textile industries rely heavily on colloid chemistry. It appears as well in the brewing of beer and in clarifying wine, making up cake-mixes and rubber tyres, extracting petroleum oil from geological deposits and converting it to motor spirit, and in the fabrication of the new ceramics for special electronic engineering applications and rocket nose-cones. The aerosols for dispensing domestic products like shaving cream and hair spray have their agricultural counterpart in the sprays for dispersing insecticides and weed-killers. On the other hand these latter techniques are also used for making defoliants and riot-control gases as well as for flame-throwers and Napalm.

The enormous contribution which colloid chemistry has made, and can still make, to technology extends also to the reduction of the harmful effects of technological development. Many of the problems listed under the heading of 'pollution' can be solved only by the use of colloid science. The clarification of water and of air are essentially colloidal problems and the specific absorptive properties of colloidal materials will no doubt be used extensively in future for concentrating (and possibly recovering) industrial products which are at present allowed to pollute the air and waterways.

1.3 Lyophilic and lyophobic sols: stability and instability

Colloids are traditionally divided into two classes called *lyophilic* ('solvent-loving') and *lyophobic* ('solvent-hating') respectively, depending on the ease with which the system can be redispersed if once it is allowed to dry out.

Lyophilic colloids can be dispersed merely by adding a suitable solvent (that is, a dispersing medium) to the dry colloid which will first swell as it takes up the liquid and will finally form a homogeneous colloidal solution. (This is what happens when making a jelly from the protein, gelatin.) A lyophobic colloid on the other hand can only be dispersed by vigorous mechanical agitation (or by the application of some other external source of energy). Most biological colloids are lyophilic in character, especially the very small ones like proteins and polysaccharides. Larger structures like chloroplasts have some lyophobic character. Many technological and agricultural materials are also lyophobic in character and it is on those systems that we will concentrate most attention.

A sol in which the particles remain separated from one another for long periods of time (of the order of days at least) is said to be **stable**. For lyophilic colloids the stability results from the fact that the solution is **thermodynamically stable** (that is the solution has a lower Gibbs free energy than the separated components); such a sol can be stable indefinitely.

For a lyophobic colloid, there is always an attractive (van der Waals) force between the particles (§2.3.5) and if they get close enough together that force will dominate so that the particles become linked together. The system can be made to appear stable for some time only if some other force is present which can reduce the chance of the particles closely approaching one another in the course of their Brownian motion. They are still **thermodynamically unstable**, however, and the barrier to coagulation is merely a kinetic one. Given enough time they will ultimately form aggregates. To say that a colloidal sol is stable then, is a relative notion: highly stable lyophobic colloids can appear homogeneous for weeks, even months but coagulation is still going on all the time, although at a very slow rate.

What other forces are involved in lyophobic systems? Although the solvent does have a role to play, the fact that the sol does not disperse spontaneously suggests that solvation effects are not sufficient to produce a stable sol. The additional repulsion can be either electrostatic or steric in origin (§2.3). If the repulsive forces are not strong enough, the particles will clump together to form what are called **flocs** or **floccules**; such a system is said to be **unstable** in the colloid sense.

Like all attempts to classify the natural world into clear-cut classes the distinction between lyophilic and lyophobic colloids is not always as simple as the above discussion suggests; some systems have some of the characteristics of each. To confuse the distinction still further, we should also note that a lyophobic colloid can be made to behave like a lyophilic one if it is coated with a sufficiently thick layer of a polymer to mask the effect of the underlying particle.

Lyophilic colloids are not obviously affected by the salt concentration of the surrounding medium until it becomes quite high. Lyophobic colloidal sols on the other hand can be markedly affected – and rendered unstable – by quite low concentrations of salt ($< 10^{-2}$ M). There is a well developed theory to describe the interaction between particles of a lyophobic colloid (Chapter 9) but the behaviour of lyophilic colloids is more difficult to describe theoretically. The reason for this is that all of the forces involved in lyophobic systems are again important for lyophilic systems but in addition there are very strong specific solvent effects which are difficult to predict.

1.4 Preparation of special colloidal suspensions

Most attempts to describe the behaviour of colloidal systems in mathematical terms assume that the particles are spherical. For larger particles (around 1 μm in size) it is possible to treat them as 'infinitely' large flat plates. (That may seem strange, but they are large in comparison with atomic dimensions and the 'infinite' flat plate needs only to be large in comparison with the distance over which the electrical or ionic forces can act, and that is rather less than 1 μm.) Sometimes they can be treated as cylinders or discs, and occasionally as spheroids. Unfortunately, real colloidal systems seldom conform to these idealizations.

Until recently most suspensions, whether obtained from natural sources or prepared in the laboratory, were of very varied particle size. Fine particles produced by grinding often vary in size over several orders of magnitude, and the naturally occurring clay minerals and oxide systems often have a similar size range. In most cases the particles are also of irregular shape, so that a theoretical description of their behaviour, except in the most general terms, is all but impossible.

Fortunately, it has now become possible to prepare suspensions of many of the most important colloids in which the particles are of highly uniform size and shape. Such systems can be used to test our theoretical predictions and, as experience and confidence grows, the insights obtained from these 'ideal' systems can be used to help explain the behaviour of the 'real' industrial or biological systems. There are also processes in which the advantages to be gained by the use of these highly uniform materials are sufficient to justify the additional labour involved in their production; but first a word about the production of the more traditional particle systems.

The procedures can be divided into two classes: **dispersion** and **condensation**, with condensation being the more important procedure for fine materials. In the dispersion methods, a sample of bulk material is broken down into the appropriate size range by grinding in a ball mill or by the strong shearing action of a very high speed stirrer (Fig. 1.1). In such processes it is usually necessary to provide a suitable **dispersing agent** in the

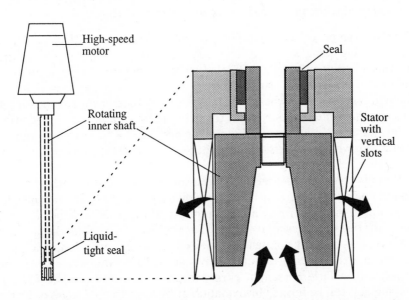

Fig. 1.1. Principle of operation of the *Ultraturrax* high-speed stirrer/homogenizer. The suspension is drawn up into the base of the instrument by the action of the rotor. It is then forced out laterally through the slots in the external tube. Because of the high velocity of the fluid, only a small volume is being processed at any one time but the laboratory device can treat a few hundred millilitres of suspension in a few minutes. All of the fluid must pass through the narrow gap between stator and rotor, where the shearing process tears aggregates or large droplets apart.

solution to prevent the small particles from aggregating together (see §1.4.2 below). A similar effect can be achieved, especially for liquid-in-liquid systems, by subjecting two liquid phases to a very high-frequency sound wave (around 20 kHz). **Ultrasonication** as it is called also requires a dispersing agent for best results. In industrial systems the high-speed stirring is sometimes assisted by the addition of another set of particles. Titanium dioxide is ground down to 200 nm size by stirring it rapidly with silica sand which itself becomes so finely divided that it coats the titania particles.

Bredig's method, in which an electric arc is struck between two wires placed under the surface of a (non-conducting) liquid is intermediate between the dispersion and condensation procedures. Presumably some of the fine metal sol so produced is formed by the explosion of the wire whilst some forms by condensation from the metal vapour.

Condensation methods are much more numerous and diverse. They may involve:

(1) dissolution and reprecipitation;
(2) condensation from the vapour; or
(3) chemical reaction to produce an insoluble product.

As an example of (1), a paraffin wax sol can easily be produced in water by dissolving some wax in ethanol and pouring the solution into a large beaker of boiling water (*CAREFUL!!*). The ethanol rapidly evaporates and an opalescent colloidal suspension of solid paraffin particles remains. Method (2) is observed whenever a mist forms by spontaneous condensation of a supersaturated vapour, as the temperature falls below the condensation point. Rapid cooling through that point gives rise to a high degree of supersaturation which favours the formation of small ($< 1\mu$m) droplets or particles.

Various chemical reactions have been routinely exploited for the production of colloidal dispersions. Most precipitation reactions can, in principle, be arranged to give a finely divided product, if the reaction is fast and the concentration of the reactants is high. The problem is to prevent the particles from aggregating in the presence of the excess salt. Barium sulphate sols can be precipitated by mixing barium thiocyanate and ammonium sulphate in the presence of a little potassium citrate to act as a dispersant (see §1.4.2). Sulphur sols can be formed by oxidation of hydrogen sulphide gas or by the oxidation of the thiosulphate ion in acid solution:

$$S_2O_3^{2-} + H_2O \rightarrow S + SO_4^{2-} + 2H^+ + 2e^-. \qquad (1.1)$$

This reaction proceeds at such a well defined rate (determined by the pH) that it is referred to as the 'clock reaction' and is often used to introduce basic kinetic ideas in elementary chemistry courses.

Silver halide sols form readily when a silver nitrate solution is treated with the relevant alkali halide, whilst some metal oxides can be produced as colloidal sols simply by boiling a solution of a suitable salt in order to induce hydrolysis. Boiling a slightly acid solution of ferric chloride produces a rich red-coloured ferric oxide sol which can be used to check the elementary properties of a colloidal suspension:

$$2Fe^{3+} + 3H_2O \rightarrow Fe_2O_3 + 6H^+ \text{ or}$$
$$Fe^{3+} + 2H_2O \rightarrow FeO(OH) + 3H^+. \tag{1.2}$$

More elaborate versions of this reaction have been exploited very effectively by Matijevic and his colleagues to produce a variety of metal oxide sols of highly uniform size and shape; these are called **monodisperse** systems.

1.4.1 Monodisperse colloids

The production of monodisperse colloidal sols has been a goal of colloid science since the beginning of this century. Perrin, in 1909, was able to prepare very small quantities of monodisperse gums and used them to provide experimental support for the kinetic theory of matter, to verify the existence of molecules, and to establish the value of the Avogadro number; for that work he received the Nobel prize for physics in 1926 (see §2.1). In the same year Theodor Svedberg received the Nobel prize for chemistry for proving that pure proteins had a well defined molecular weight (that is were

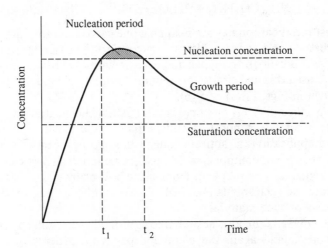

Fig. 1.2. The La Mer diagram. Producing a monodisperse sol by confining the formation of nuclei to a short period (t_1 to t_2) and allowing growth to occur only on those nuclei already formed ($t > t_2$).

monodisperse); that knowledge was crucial to the development of modern biochemistry.

The principle involved in the formation of monodisperse systems is illustrated in Fig. 1.2 (often called a La Mer diagram). Some means must be found to produce a gradual increase in the concentration of the required material in solution. Precipitation usually does not occur as soon as the concentration reaches the saturation value because the molecules or ions have no existing crystal structure on which to build (that is no **template**). When the concentration reaches a certain degree of supersaturation, the driving force for formation of the solid is sufficient to induce the formation of the tiny nuclei which will ultimately grow into the final particles; this is the **nucleation concentration**.

Formation of the nuclei reduces the solution concentration and, from that point on, the aim is to maintain the solution concentration between the nucleation and the saturation concentration. If that can be done there will be no new nuclei formed and all subsequent growth occurs on those formed in the first burst of nucleation. It is this which ensures the uniformity of the particle size. Zaiser and La Mer (1948) used this technique to produce highly monodisperse sulphur sols from acid thiosulphate solutions (eqn 1.1).

One problem is to find a method of feeding more of the desired material into the system at just the right rate to make up for particle growth and maintain the desired solution concentration. That is usually done by controlling the concentration of one of the ions involved in the precipitation. For example, one can use the slow hydrolysis of urea to form the carbonate ion:

$$NH_2 \cdot CO \cdot NH_2 + 2H_2O \rightarrow CO_3{}^{2-} + 2NH_4{}^+. \qquad (1.3)$$

Since most metal carbonates are insoluble one can obtain the carbonate or basic carbonate and convert this subsequently to the oxide by heating (or **calcining** as it is called). Matijevic and his associates (see Matijevic and Bell 1973) have used this and similar procedures to obtain colloidal sols such as those shown in Figs. 1.3(a) and (b).

Colloidal silica (SiO_2) is an important industrial chemical, used as a filler for paint, a catalyst support, a rubber reinforcing agent, and for such specialized applications as antireflecting agents and for encapsulating compounds for electronic components. Monodisperse spherical particles of silica (silicon dioxide) can be prepared from silicic acid or by hydrolysis of ethyl orthosilicate and such sols can be used to make synthetic opal, since natural opal consists of such material.

Organic polymers can also be prepared as monodisperse sols, by emulsion polymerization, and again the method depends on generating a limited number of nuclei and building on them to produce the final product. Indeed, these were the first such sols to be prepared in large quantities almost at will. We will describe that method in some detail when we have introduced a few new ideas on the behaviour of simpler systems (§1.4.5).

Fig. 1.3. Monodisperse inorganic colloids. (a) Zinc sulphide (sphalerite); (b) cadmium carbonate. (Photograph courtesy Professor Egon Matijevic, Clarkson University, New York.)

Instead of forming nuclei in this way it is also possible to prepare a solution in the concentration range between saturation and nucleation and then introduce a few tiny seed crystals, either of the desired product or of another material of the same or similar crystal habit. The initial small size differences are gradually eliminated as the particles grow. To be successful one must not allow other nuclei to form during the growth stage.

1.4.2 Surfactants

Surfactant is a contraction of **surface active agent** and refers to a class of chemical compounds known technically as **amphiphiles** (from two Greek words meaning that they are not certain what they like)†. The molecules of such compounds consist of two regions of very different characteristics: one part is polar (either a dipole or a charged group) and the other is non-polar (usually a hydrocarbon or halocarbon chain) (Fig. 1.4). Since each part has very different solubility properties, the molecules have limited solubility in any solvent and tend to accumulate at the interface between two phases, where the polar part can immerse itself into the more polar phase, and the non-polar part can do likewise.

The best known examples are **soaps**, which are usually sodium or potassium salts of organic (fatty) acids, like oleic, palmitic, or stearic acid. Sodium stearate ($CH_3(CH_2)_{16}COO^-Na^+$) is the basic ingredient of common soap whilst potassium oleate ($CH_3(CH_2)_7CH=CH(CH_2)_7COO^-K^+$) finds uses in softer, higher-quality soaps. (The double bond makes the chain more rigid, interferes with the lateral interaction between the molecules and leads to a lower melting point and less rigid behaviour at interfaces. Recall the behaviour of 'polyunsaturated' vegetable oils used as substitutes for butter.) Similar salts of aluminium and some heavy metals are used to make lubricating greases.

At low concentrations surfactants form true solutions but some molecules will begin to adsorb onto the walls of the containing vessel and/or the air/solution interface (Fig. 1.5) because, in that way, it is usually possible for the two distinct parts of the molecule to find a more favourable environment. As the concentration rises, the adsorption increases and a point is reached at which adsorption becomes even more favourable because the adsorbed molecules can begin to interact laterally with one another through the mutual attraction of their hydrocarbon chains. The surfaces then become covered with a monolayer of the surfactant.

At about this stage, further dissolution in the normal way ceases, but now a new process becomes possible. The molecules in the solution begin to aggregate into what are called **micelles**, containing some 50–100 individuals. This process is a cooperative one and it occurs at a fairly precisely defined

---

† Sometimes called 'amphipathic' compounds, meaning they don't know what they hate.

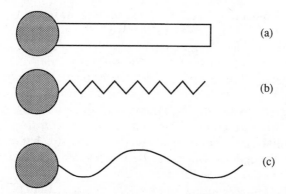

(a)

(b)

(c)

Fig. 1.4. Representations of an amphiphile or surfactant molecule. The round *head group*, is polar and the *tail* is non-polar. The tail is flexible within the limits imposed by rotation around the C–C bond. The cross-section of the hydrocarbon chain is about $0.2 \, nm^2$ when fully extended, and this is comparable to the head group size for $-OH$, and $-NH_2$, but smaller than for $-SO_4^-$. The representations here emphasize: (a) space filling, (b) number of carbons in the chain, and (c) flexibility.

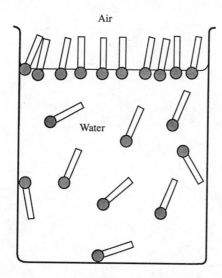

Air

Water

Fig. 1.5. Schematic arrangement of surfactant molecules in a solution at low concentration. Some molecules are preferentially adsorbed at all available surfaces. (Counterions are not shown.)

concentration, called the **critical micellar concentration (c.m.c.)** which can be identified from the fact that the equilibrium and transport properties of the solution (like optical density and electrical conduction) are affected by the aggregation process.

The exact number of molecules in a micelle and the resulting shape depend on a number of factors, like the relative sizes of the two parts of the molecule, the presence or absence of double bonds and aromatic rings, temperature, and the concentration of surfactant and other electrolyte. Sodium dodecyl sulphate (SDS) micelles are spherical in shape when first formed (Fig. 1.6) but as they grow in size with increasing surfactant concentration, they tend to become less symmetrical and ultimately form discs and a variety of other structures. Note that in the case of a soap micelle, there is a substantial charge developed on the aggregate, but part of this is balanced by the cations (called **counterions**) which tend to remain closely associated with the micelle surface.

As a result of micelle formation, a vastly greater amount of the surfactant is able to 'dissolve' and the micelles act as reservoirs of surfactant which can readily supply molecules to any new surface which becomes available. This is why the surfactants are so useful as cleaning agents and stabilizers and why their solutions also produce a very stable foam if they are shaken.

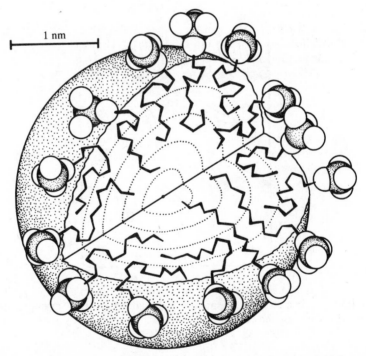

1 nm

Fig. 1.6. A sodium dodecyl sulphate micelle. This is the picture which emerges from a detailed statistical mechanical calculation of the likely structure. (Drawn by Professor J. N. Israelachvili from calculations by Dr D. W. R. Gruen.)

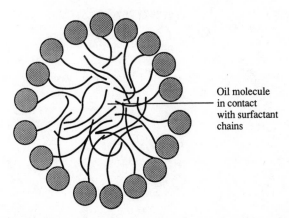

Oil molecule in contact with surfactant chains

Fig. 1.7. A surfactant micelle swollen by the presence of some solubilized oil. The molecules of oil are all in intimate contact with surfactant hydrocarbon chains. Organic molecules of any kind may be solubilized in this way, if they are soluble in liquid paraffin.

Another interesting consequence of micelle formation is the phenomenon of **solubilization**: an organic compound which would normally be insoluble in water can be 'dissolved' in a surfactant solution because it can move into the oily interior of the micelle (Fig. 1.7). This is one of the mechanisms by which a soap acts as a cleaning agent or **detergent** (Chapter 10); the other is its ability to adsorb on the surface of an oil droplet and so **stabilize** an oil-in-water (O/W) emulsion (see §1.4.3).

At higher concentrations, the amphiphiles form a variety of structures of which an interesting example is the **liquid crystal**, in which the molecules are ordered in some directions but able to move, more or less freely, in others. In some cases, the molecular orientation can be changed by applying an electric field to the system and in that case the substance may be used to provide a 'liquid-crystal display' (LCD) such as is used in meters, calculators, and lap-top computer screens.

1.4.3 Emulsions

An emulsion is a dispersion of a liquid in another liquid in which it is essentially insoluble. Emulsions are usually produced by dispersion rather than condensation methods (§1.4), because dispersion is much easier for a fluid than for a solid. The formation of a large area of interface between two phases normally requires the expenditure of a significant amount of energy (vigorous stirring or shaking), but this can be dramatically reduced in the presence of a surfactant, which may act as a **dispersing or stabilizing agent**.

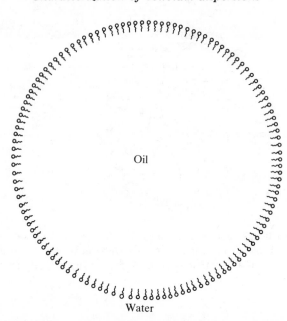

Fig. 1.8.   An oil droplet in water; the drop is prevented from contacting and coalescing with other droplets by the presence of an adsorbed layer of surfactant. The head group of the surfactant molecule would commonly be a carboxyl ($-COO^-$) or sulphate ($-OSO_3^-$) group. Repulsion forces between the headgroups are reduced by the presence of positive counterions (§1.4.2).

The surfactant adsorbs at the interface between the oil and water so that the droplets of oil become charged (usually negatively) (Fig. 1.8). The electrical charge produces a repulsive force between approaching droplets, and this prevents them from making close contact during a collision and coalescing[†] together.

Some systems are so readily dispersible they are referred to as **spontaneously emulsifiable**, though they usually require some input of mechanical energy (a little shaking). Sometimes, though, the energy required to form the new surface is provided by chemical or physical reactions within the system itself. (See Exercise 1.4.8.)

A special case of 'spontaneous' emulsion' formation occurs in some systems which form very small droplets (around 10 nm radius); these are referred to as **microemulsions**. The surface energy or interfacial tension[‡]

[†] Coalescence refers to the process by which two droplets, when they collide, reform into a single droplet.
[‡] Surface energy is measured per unit area ($J\ m^{-2}$). The same surface property can be represented as a force per unit length ($N\ m^{-1}$); it is then called surface (or interfacial) tension. Thus, the surface tension, γ, of pure water at 25°C is $72\ mN\ m^{-1}$ or $72\ mJ\ m^{-2}$. The link between these two will be established in §5.2.1.

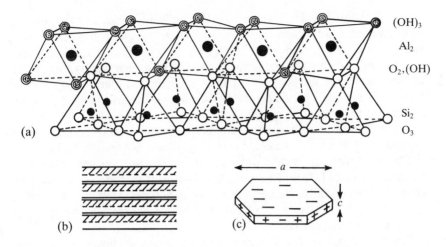

$(OH)_3$

Al_2

$O_2,(OH)$

Si_2

O_3

Fig. 1.9. (a) A sketch of the 'ideal' kaolinite layer structure $(Al(OH)_2)_2 \cdot O.(SiO_2)_2$. One hydroxyl ion is situated within the hexagonal ring of apical, tetrahedral oxygens and there are three others in the uppermost plane of the octahedral sheet. The two sheets combined make up the kaolinite layer. (b) Simplified schematic diagram of a kaolinite crystal. Note that the upper and lower cleavage faces of the perfect crystal would be different. A typical crystal would contain about a hundred or so such layers. (c) A typical kaolinite crystal of aspect ratio (a/c) about 10. Note the negative charges on the cleavage faces or *basal planes* (perpendicular to the c axis of the crystal) and positive charges around the edges. The latter are eliminated at pHs above about 7.

in those cases must be very small, since very large surface areas may be involved. The droplets are so small that they do not scatter light and the emulsion may therefore be quite transparent, whereas normal emulsions are typically opaque white (like milk) because they scatter light so very well. At the lower end of the size range, an oil-in-water microemulsion grades almost imperceptibly into the solubilized oil-in-micelle structure of Fig. 1.7. Microemulsion systems are being studied for their potential in tertiary oil recovery, among other things.

1.4.4 Clay minerals

The clay minerals are distributed widely in nature and have been used by man in a variety of ways since the Stone Age. The term **clay** is used in agriculture and soil science to refer to any soil particle of radius less than about $2\,\mu m$ but the bulk of that material in a normal soil consists of particles of one of the **aluminosilicates** which are called the **clay minerals**. They owe their properties to the fact that the silicon atom can bind to oxygen atoms to form extensive thin flat sheets in which each silicon is surrounded by four oxygens in a tetrahedral arrangement. These silica sheets are able to bond to other flat sheets of aluminium oxide (AlO(OH)) in which each aluminium atom is surrounded by an octahedral arrangement of oxygen and hydroxyls. The double sheet structure can be stacked in layers bound together by hydrogen bonds. This is the basic structure of the mineral **kaolinite** (Fig. 1.9), which is referred to as a 1:1 layer silicate.

An important feature of many colloidal clay minerals is the fact that the mineral crystal carries a negative charge, which is thought to be due to defects in the crystal lattice: some of the silicon (Si^{4+}) atoms are replaced by aluminium (Al^{3+}) and each such substitution† produces one negative charge. Some of those charges are probably compensated by other defects in the lattice but many of them remain to produce a net negative charge on the outside of the crystal.

For most purposes the kaolinite crystal behaves as shown in Fig. 1.9(c). The negative charge on the cleavage faces is balanced by the presence of counterions (§1.4.2) from the solution. That negative charge is an important characteristic of the clay, called its **cation exchange capacity (c.e.c.)**. The c.e.c. of a clay mineral may be determined by taking a known mass of clay, rinsing it several times with a solution of an ammonium salt, until all the charges are balanced by ammonium ions, washing it with water to remove the excess salt, and then determining how many ammonium ions are present (usually by treating the sample with NaOH, distilling off the ammonia into an acid and titrating the remaining acid). (Exercise 1.4.9.)

Another large class of clay minerals is based on a 2:1 layer structure in which an alumina layer is sandwiched between two silica layers (Fig. 1.10). Since the outer layers in this case are all oxygen atoms there is no opportunity for hydrogen bonding between the triple layers. They are held together only by van der Waals forces and the crystals can therefore be very easily cleaved between the contacting silica layers. The basic mineral is called **pyrophyllite** but the more interesting material from our viewpoint is produced when some (about a quarter) of the silicon atoms are replaced by aluminium to produce the mineral **muscovite** or **white mica**. Again this produces one negative charge on the crystal lattice for each substitution.

The negative charge is normally balanced by the presence of potassium ions which can fit snugly into the hexagonal holes in the silica surface. The resulting strong electrostatic attraction holds the sheets together so that the mica crystal is quite stable in contact with water. Mica crystals can be quite large (many centimetres across) and are relatively easy to cleave in air or vacuum. Good specimens can be cleaved to yield pieces which are atomically smooth on a macroscopic scale; that is a most unusual phenomenon and such

† This is called isomorphous (same shape) substitution because the replacing and the replaced ion must be of about the same size so that the lattice is not too badly strained.

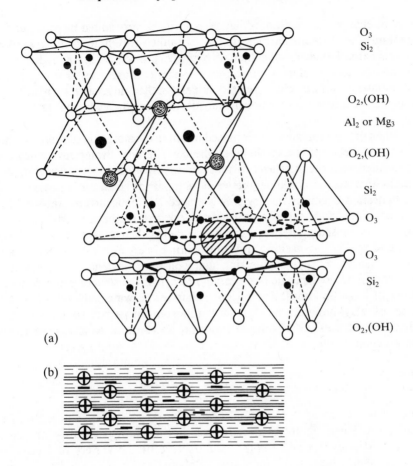

O_3
Si_2

$O_2,(OH)$

Al_2 or Mg_3

$O_2,(OH)$

Si_2

O_3

O_3

Si_2

$O_2,(OH)$

(a)

(b)

Fig. 1.10. (a) A sketch of the ideal 2:1 layer lattice aluminosilicate. The large circle shows the preferred position of the counterion required to balance the crystal charge after isomorphous substitution. (b) Schematic diagram of the white mica structure. The potassium ions are shown + and they balance the negative charges in the silica layers caused by the substitution of about a quarter of the silicon atoms by aluminium. The 'ideal' formula is $K^+[(AlO(OH))_2(AlSi_3O_8)]^-$.

specimens have proven to be invaluable for the study of the nature of the forces between colloidal particles (see §9.4.3).

If the pyrophyllite structure is modified by the replacement of about one in six of the aluminium ions by magnesium, then again there is a negative charge generated for each substitution and cations must balance that charge within each sheet. In this case, however, the mineral (called **montmorillonite**) forms very small crystals in which a variety of ions may act to balance the crystal charge. The cation exchange capacity is very much larger than for kaolinite, because the surface area is very much (often 20–30 times) larger in this case. The material called **Wyoming bentonite**, which is widely used in industry as a catalytic support and in drilling muds, consists of a mixture of montmorillonite and a related mineral called **beidellite** in which the aluminium ion is isomorphously substituted for silicon in the tetrahedral layers.

When dry montmorillonite is placed in a moist atmosphere it is able to take up water between the triple-layer sheets. The first water molecules are associated with the interlayer cations and the hydration continues until two layers of water are contained between each pair of sheets. The same degree of hydration occurs if the dry clay is placed in a concentrated (about 1 M) salt solution. If the salt concentration is lowered, so that the activity of the water is increased, more water can penetrate between the sheets and the clay mineral swells. The swelling and contraction of clay minerals has important consequences in civil engineering and construction since the force and pressure generated can be large enough to affect the structure and foundations of large buildings. The cation exchange and water adsorption behaviour of clay minerals is also very important in agriculture since they determine the ability of soils to transport and supply water and nutrients to growing plants.

1.4.5 Emulsion polymerization

This process is used to supply most of the world's growing requirement for polymers. Huge quantities are now used in the paint, packaging, and construction industries and for the manufacture of the variety of consumer goods we now fabricate from plastics. Polymerization is an exothermic process and the early methods of **bulk polymerization**, in which the polymer forms from a monomer in solution in an organic solvent, were difficult to control and often resulted in a highly variable, even useless, product. The emulsion method allows the heat output to be controlled and the product is also a free-flowing aqueous dispersion from which the polymer may readily be separated, if necessary.

In emulsion polymerization, the monomer (for example styrene) is made up as an emulsion (§1.4.3) in water, usually in the presence of a surfactant (§1.4.2) to act as a stabilizer. The initiator is dissolved in the aqueous phase, which will also contain a little monomer, and the initial polymerization begins there. The function of the initiator is to provide a source of **free radicals**, and a typical water-soluble example is the persulphate ion:

$$S_2O_8{}^{2-} \rightarrow 2SO_4{}^{\cdot-} \tag{1.4}$$

The sulphate radical†, which we will represent as $I\cdot$ can react with a molecule of the monomer (M) to produce a new radical:

$$I^{\cdot} + M \rightarrow M^{\cdot} \qquad (1.5)$$

in what is called the **initiation** step. Monomer molecules then add on to this free radical in a **propagation** step:

$$M^{\cdot} + M \rightarrow M_2^{\cdot}$$
$$M_r^{\cdot} + M \rightarrow M_{r+1}^{\cdot} \qquad (1.6)$$

and this process continues until the number of units in the chain (r) is several hundred or even several thousand. The process stops when the free radical is destroyed in one way or another, in a **termination** reaction, for example:

$$M_r^{\cdot} + M_n^{\cdot} \rightarrow M_{r+n} \qquad (1.7)$$

in which the two unpaired electrons become paired.

The stabilizing surfactant will be present partly in solution and partly in the form of micelles. The details of the subsequent process are still the subject of active research, but it seems that, after an initial formation of nuclei, the tiny polymer particles (called **oligomers**) enter a surfactant micelle, which is able to **solubilize** them, and there they grow by using up monomer which is continually diffusing into the micelle from the surrounding solution. The droplets of monomer act as reservoirs to provide a supply to the growing polymer chains inside the micelles (Fig. 1.11).

Monodisperse samples of polystyrene and poly(methyl methacrylate) (PMMA) have been prepared by this method and they have been widely used as models for studying the behaviour of suspensions of perfectly spherical particles. They are negatively charged due to the presence of surface groups which originate from the initiator, or the dispersing agents used to stabilize the initial emulsion or by oxidation of the monomer.

Polymeric spheres with varying numbers of acidic ($-COO^-$) and basic ($-NH_2$) surface groups can also be prepared (Homola and James 1977). The carboxyl groups can be either neutral or negatively charged depending on the pH, and the amine groups can be either neutral or positive. As the pH changes from acid to basic the charge moves from positive through zero to negative just as it does for a protein:

$$\text{Low pH: } -NH_2 + H^+ \rightarrow -NH_3^+,$$
$$\text{High pH: } -COOH + OH^- \rightarrow -COO^- + H_2O. \qquad (1.8)$$

---

† A free radical is a neutral or charged atom or molecule in which one of the electrons remains unpaired. That electron is represented by the superscript $\cdot$.

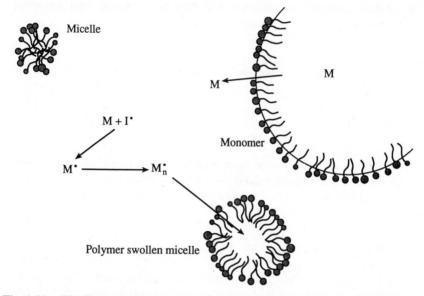

Fig. 1.11. The likely mechanism of emulsion polymerization. The large droplets of monomer (M) are stabilized by an added surfactant or by surface-active agents generated in the early reaction stages. Some monomer dissolves in the water and the initial polymerization begins there $(M + I \cdot \rightarrow M \cdot \rightarrow M_n \cdot)$. The tiny polymeric nucleus diffuses to a micelle and there continues to grow, using up monomer which is continually replenished from the emulsion droplets, by diffusion through the solution.

The particular pH at which the negative and positive charges are balanced (so there is no net charge on the colloid) is called the **point of zero charge** and the pH at which it occurs will depend upon the relative numbers of acidic and basic groups on the surface of the polymer latex (Exercise 1.4.10). These are referred to as **amphoteric** or, more correctly, **zwitterionic** latex systems.

Exercises

1.4.1   Under what circumstances would you expect the particles of a paraffin wax sol, produced as described in §1.4 to be approximately spherical in shape?

1.4.2   Why does a stable foam form when a soap solution is shaken?

1.4.3   How does the presence of micelles assist in the detergent (cleaning) action of a soap solution?

1.4.4   A typical recipe for making a latex (van den Esker and Peiper 1975) uses 25 g of styrene ($C_6H_5CH=CH_2$) in 25 ml of water containing 1 g of sodium dodecyl sulphate (SDS) ($(CH_3(CH_2)_{11}SO_4^- Na^+$) as stabilizer. Assume that the styrene (density 0.9 g ml^{-1}) forms emulsion drops of average radius 3 μm and that the whole surface of these droplets is covered with SDS which occupies $0.5 \, nm^2$ per molecule. How much SDS is used in the process?

1.4.5 Treating the same system as in Exercise 1.4.4, take the critical micelle concentration (c.m.c) of SDS as 8 mM and assume that all SDS in excess of that amount is either on the surface of the emulsion droplets (as calculated in Exercise 1.4.4) or in the form of micelles. How many micelles are there per ml assuming that there are 50 molecules per micelle? How many SDS molecules in the bulk solution? How many emulsion droplets? Is a free radical more likely to collide with an emulsion droplet or with a micelle? (Take account of both the number and the cross-sectional area of the target.)

1.4.6 The solubility of styrene monomer in water is $0.5 \, g \, l^{-1}$. How many styrene molecules will there be per ml of solution? Compare the result with that in Exercise 1.4.5 to determine whether it is more likely that a free radical will collide with a dissolved styrene molecule (of cross-section $0.5 \, nm^2$) rather than with an emulsion drop or a micelle.

1.4.7 Speculate on why (polyunsaturated) 'margarine' spreads more easily than butter.

1.4.8 Make up a solution containing 14% ethanol in toluene and pour it carefully onto the top of a layer of water in a beaker. Observe the interface between the two layers. The milkiness is due to formation of an emulsion of water in the toluene. (*Careful: toluene vapour is toxic so this should be done in a well ventilated place, like a fume cupboard.*) Davies and Rideal (1963) give a detailed explanation (with pictures).

1.4.9 (a) A kaolinite sample (2 g) is saturated with ammonium ion to determine the cation exchange capacity (c.e.c.). The sample is then treated with NaOH and the resulting ammonia distilled into 25 ml of 0.0300 M HCl. The acid is then found to require 32 ml of 0.0215 M NaOH for neutralization. Estimate the c.e.c. in millimoles (of monovalent ion) per 100 g of clay.

(b) If the particles are assumed to be disc shaped (Fig. 1.9(c)) with a radius of 700 nm and a thickness of 50 nm, calculate the distance between the centres of the negative charges (assuming they are confined to the basal (cleavage) faces).

1.4.10 (a) Consider a zwitterionic (or amphoteric) polymer latex system with particles of radius 100 nm. The density of amine groups is 1 per $20 \, nm^2$ of particle surface and the density of carboxyl groups is twice that. The particle concentration is $10^{12} \, cm^{-3}$. The pK_a of the amine is 8 and that of the carboxyl is 5. What is the concentration of amine groups (moles per litre)? Estimate the positive, negative, and net charge on the particle at pH 4, 6, 8, and 10.

(b) Repeat the calculation with a density of one carboxyl per $20 \, nm^2$ and one amine per $10 \, nm^2$.

(c) Estimate, to within 0.5 pH units, the p.z.c. in the two cases. (*Hint:* The pK_a of the amine refers to the acid ionization reaction and implies that, for the ideal systems assumed here: $K_a = 10^{-8} = [-NH_2][H^+]/[-NH_3^+]$. Concentrations of the surface species may be expressed per unit area of solid or per litre of solution for the purpose of the exercise.)

1.5 Maintaining clean surfaces

One of the most important problems encountered by colloid scientists, especially when dealing with lyophobic colloids, is the fact that the properties of colloids can be dramatically influenced by the presence of quite small amounts of certain substances (particularly surfactants, multivalent ions, and polymeric materials). Even in industrial situations where it is difficult to deal with clean systems, it frequently happens that a suitable product is obtained only when one or more troublesome impurities is rigorously excluded from the system.

There are three separate impurity problems: the composition of the particles themselves; the possibility of a (slow) chemical reaction occurring on the surface to gradually alter its nature; and the possibility of adsorption of some material from the surrounding solutions at some stage in preparation, treatment, and storage. Once the system has been cleaned up it is necessary to ensure that all solutions with which it will come in contact are free of contaminants. One must be particularly careful in situations where the particle concentration is small, so that the available surface is small, for in such cases a minute amount of a contaminant may be sufficient to cover the entire surface and completely alter the properties of the system. (Exercise 1.5.1.)

1.5.1 Particle composition

Colloidal particles of natural origin, such as the clay minerals, may vary significantly in composition depending on the source. The important commercial clays, like kaolinite, are often compared to specimen samples of those clays mined from particular sites where the clay occurs in a near-pure form. Thus Georgia kaolin, Fithian illite, and Wyoming bentonite are to the clay mineralogist something like AR reagents for the analytical chemist. Though it has proved possible to synthesize small quantities of some of the clays, that is still a difficult exercise. For the most part we must be content with discovering the 'purest' naturally occurring deposits. For biological samples separation techniques have now reached a high degree of sophistication, but to discuss them would take us too far into the realm of biochemistry.

For synthetic systems, like the metal sols, oxides, sulphides, and the silver halides, one might think that there were few problems, but that is not so. The bulk properties may be well known but the details of the surface structure may require considerable study, even for simple systems. For example, when one attempts to prepare a monodisperse sample of silver iodide, it seems that the more monodisperse it becomes, the more variable is the structure of the surface layers. Fortunately, we now have a variety of physical analytical techniques with which to study the surfaces of such systems (see Adamson 1990 for an introduction to some of the newer techniques).

1.5.2 Chemical reaction at the surface

The most common problem encountered is surface oxidation. One can rigorously exclude oxygen during the preparation procedure, using the usual dry-box techniques, but it may be more difficult to prevent the slow oxidation which occurs on prolonged storage. Keeping the dry material in an inert atmosphere should suffice, but if this is not possible one must either make the material up afresh for each application or develop a suitable method for removing surface contaminating layers. Oxide layers will usually dissolve in acid, whilst for metal sulphides the product is often a thiosulphate, or free sulphur. Once the product is known it is usually not difficult to devise a suitable etching or washing procedure to remove it.

1.5.3 Adventitious impurities

The most pervasive problem is that caused by the presence of chance impurities derived from the preparation or storage procedures. Grinding in a ball mill (with metal balls) may introduce polyvalent metallic ions, but these can be fairly easily removed by ion exchange: the colloid is washed with a fairly concentrated solution of a simple electrolyte like KCl (1 M) at pH 3 and then with water to remove the excess salt.

More stubborn problems are caused by the adsorption of polymeric contaminants: organic polymers from the membranes used for filtration and dialysis (§1.6.1) and inorganic oligomers like silica derived from the walls of glass containers. The organic polymers are particularly difficult to remove from polymer latex systems, with which they have a high affinity.

Glass is usually regarded as the most satisfactory container for most chemical systems. Although borosilicate glass is rather unreactive towards most chemicals (except fluoride) it does react with alkali. Storing a colloidal suspension in a glass container, especially at pH > 7 will almost invariably result in its acquiring a coating of silica. Biological samples are not so susceptible but inorganic (especially oxide) sols are. Using plastic containers is not always an ideal solution. The low molecular weight organic substances used as 'plasticizers' are often surface active and can slowly diffuse from the container walls to coat the colloid. Storage under dry conditions is obviously wiser, if the sol can be readily redispersed.

Exercise

1.5.1 A dilute colloidal dispersion contains $0.01\,\mathrm{g\,l^{-1}}$ of solid particles with an average radius of 500 nm and a density of $2.6\,\mathrm{g\,cm^{-3}}$. Estimate the surface area of solid (in cm^2 per litre). Note that the result is comparable with the surface area of the container. A contaminant of molar mass 320 adsorbs on the particle surfaces and occupies an area of $1.1\ nm^2$ per molecule. What mass is required to cover half the surface of the solid particles in 1 l of dispersion?

1.6 Purification procedures

Colloidal purification procedures depend upon the fact that the impurity is usually in equilibrium with material in the solution phase, and is not irreversibly adsorbed. The procedures depend upon either:

(1) successive dilution of the impurity in the solution phase; or
(2) competitive adsorption of the impurity onto another surface.

In successive dilution methods the particles are brought into contact with a large volume of liquid and allowed to equilibrate; the two are then separated and the process is repeated with a fresh sample of liquid. This 'batchwise' procedure can be replaced by a continuous one in some cases.

The competitive adsorption procedure can be used only when the competing surface can be readily separated from the colloid, either because of its physical form (for example activated carbon cloth or carbon fibres) or particle size (for example ion exchange resins).

1.6.1 Solute dilution procedures

The only problem with these procedures is the separation of the liquid from the colloid at each step. This can be done by sedimentation (either under gravity or in a centrifuge) or by filtration using a membrane that is permeable to the liquid and the impurity but not the colloid.

(a) Sedimentation and centrifugation Sedimentation will be discussed in more detail in the next chapter. We need only note here that particles of radius greater than 1 μm and relative density greater than about 2.5 will settle under gravity in a reasonable time. They will certainly sediment rapidly if the particles have aggregated together to form **flocs**. The rapid settling of flocs is often used in the purification of a colloid, especially a clay mineral system. Natural clays often have a variety of cations adsorbed on their surfaces to satisfy the negative crystal charge (§1.4.4). To generate a more homogeneous material we need to replace these with a single ion type and this is done by washing the clay with a reasonably concentrated (0.1–1 M) solution of a simple 1:1 electrolyte like NaCl†. In such solutions clay

---

† This is usually done at pH 3 so that adsorbed aluminium ion (Al^{3+}) will be removed. At higher pHs the Al is present as an oxy-hydroxy oligomer which is strongly hydrogen bonded to the surface and difficult to dislodge.

minerals will normally remain flocculated since they are lyophobic colloids (§1.3); the flocs settle quite quickly so that the washing process can be readily accomplished by settling under gravity and decanting the supernatant liquid.

More usually it is necessary to use centrifugation. One potential problem here is that, for small pariticles of low density, the centrifugation speeds may be so high that the initial sediment becomes very highly compressed and is very difficult to redisperse. It is often necessary to compromise somewhat on the settling time, by spinning more slowly, to avoid producing a sediment which will not redisperse at all. Effective washing can only be accomplished if the solid can be completely redispersed after each centrifugation.

The normal centrifuge head may be replaced by a continuous head for processing large volumes of dilute suspension. In this case the solution flows from a reservoir into the spinning head and the supernatant liquid can flow out of the head whilst it is spinning. The capacity is then limited only by the volume of the head available for storing the sediment.

(b) Dialysis This method was the basis of the original distinction between colloids and true solutions, as proposed by Thomas Graham in the 1860s. The colloid is retained on one side of a membrane (often in the form of a bag) which is bathed in the wash solution. The pores of the membrane must be large enough to allow passage of the solvent and small solute molecules, but not the colloid. The membrane may be said to be **semipermeable**, but that term is usually reserved for osmotic membranes which allow *only the solvent* to get through.

The usual membranes are made from reprecipitated cellulose (called 'cellophane' or 'cuprophane') in which the rather irregular pores are formed by a meshwork of fibrils. The material comes in sheet form or, more usually, as a tube which can readily be cut and tied to make a convenient bag. Before use the membrane material must be boiled in several changes of distilled water to remove soluble impurities.

Dialysis is a diffusive process in which ions and other small molecules move through the membrane as a consequence of the concentration difference on either side. If the solutions on each side are well mixed, the rate of dialysis for a particular species $(- dC_i/dt)$ is proportional to the concentration difference across the membrane, and the membrane area, A, and inversely proportional to the thickness, d:

$$dC_i/dt = - (KA/d)(C_i - C_o) \qquad (1.9)$$

where subscripts i and o refer to inside and outside the dialysis bag and t is the time; K is a constant which measures the permeability of the membrane. The process can be speeded up by using thin (sufficiently strong) membranes of large area, maintaining C_o small by frequent changes of the external bathing solution, or by using a continuous-flow system.

The driving force for any species across the membrane is the difference in activity or chemical potential between the inside and the outside. For the colloid this is of little importance since we assume that, no matter how high the driving force, the membrane will be strong enough to hold it in. For the solvent, the lowering of chemical potential produced by the dissolved solutes in the bag causes a driving force (the osmotic pressure) into the bag. If the salt concentration inside the bag is high the volume of water flowing in may create a problem. In any case that inflow will oppose the outward movement of solute which is the primary purpose of the dialysis process. Fortunately, the retarding effect of osmotic inflow is highest in the early stages, when the salt concentration is highest and the rate of outward diffusion of the solutes is also highest. Obviously, good mixing of both regions, especially the inside suspension, speeds the attainment of equilibrium.

(c) Filtration This refers to any process in which the suspended or dissolved material is separated from a fluid as the fluid (liquid or gas) passes through a porous material. The essential distinction between this and dialysis is that, in filtration, the fluid must flow bodily through the porous material, rather than moving by a molecular diffusion mechanism as it does in osmosis and dialysis. As the fluid moves it carries the suspended or dissolved substances with it (in a process called **convection** or **convective transport**) until the suspended matter is caught in the pores. In some cases the suspended matter cannot enter the pores at all but forms a cake (the **filter cake**) on top of the porous medium or filter. In many important filtration processes, however, such as the purification of domestic drinking water, the colloidal material is filtered out by passing the (dilute) suspension through a bed of sand (**deep-bed filtration**), usually after flocculation of the colloidal impurities.

(d) Ultrafiltration Ultrafiltration is the extension of the filtration process to very small (that is colloidal) particles. This necessarily requires the use of membranes with a very small pore radius, and is usually assisted by applying pressure. It should be distinguished from the very similar process of **reverse osmosis**† which can be used to remove dissolved impurities; in that case the solvent moves by diffusion, whereas in ultrafiltration the solvent still moves by bulk (viscous) flow.

It is now possible to obtain membranes with precisely defined pore size. 'Nucleopore' membranes, for example, are made by irradiating a suitable piece of material (mica or a polymer) with ionizing radiation. The tracks left

---

† Reverse osmosis: if a solution is separated from the pure solvent by a semipermeable membrane (§1.6.1) and the solution is subjected to a pressure in excess of the osmotic pressure of the solution, then essentially pure solvent can be forced through the membrane, leaving the solute behind. The process is used in desalination.

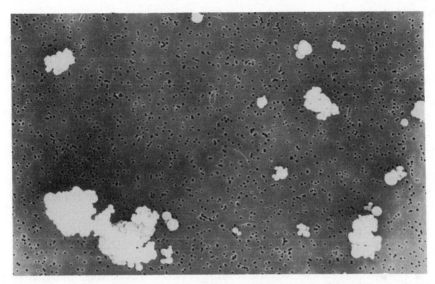

Fig. 1.12. Transmission electron micrograph (TEM) of a 'Nucleopore' membrane with pore radius 0.4 μm and some poly-(vinyl chloride) latex particles collected on it. (Courtesy of A. Rogers, University of Sydney.)

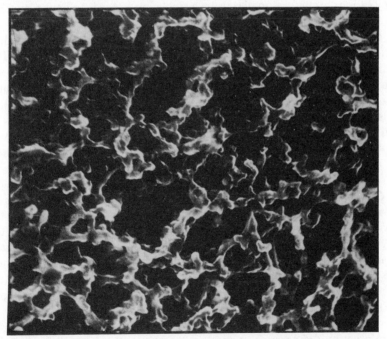

Fig. 1.13. Scanning electron micrograph (SEM) of a 'Millipore' membrane, in which the pores are produced by a cross-linked fibre matrix.

by the radiation weaken the crystal and can be etched by a suitable solvent to produce channels of regular cross-section (Fig. 1.12). Note, however, that the area available for flow through such membranes is strictly limited. A much more open structure, although with a less well defined pore size, is provided by the 'Millipore' type membrane shown in Fig. 1.13.

In some membrane filters, the membrane is formed into a hollow fibre. The solution to be filtered is forced down the lumen of the fibre and the clear filtrate moves radially out through the fibre walls. The fibres are arranged in a bundle and high filtration rates are made possible because the flow of the fluid down the tube prevents the build-up of a filter cake on the inside of the tube. This is what is called **cross-flow filtration** because the filtration occurs in a direction at right angles to the flow. It results in much improved performance in most cases.

1.6.2   Competitive procedures

(a) Ion exchange   The surface of most colloids is electrically charged, by mechanisms which we will discuss in Chapter 7. Those charges are neutralized by the adsorption of ions of opposite sign on the external faces of the colloid crystals. We have mentioned already the need to replace the mixed array of ions on natural materials by a single ion if one is to have a material of reproducible properties (§1.6.1(a)).

The same process is used for polymer latex particles, but instead of washing with a concentrated salt solution, the exchange is accomplished by placing the latex colloid in contact with a slurry of ion exchange beads loaded with the ion required. The beads must have a much higher capacity than the colloid so that when equilibrium has been attained there has been little transfer from them to the latex. Since latex systems do not normally have a high surface charge that is not difficult to arrange. The method is, however, not without problems since it is hard to ensure that the resin beads are totally free of any potential contaminant.

(b) Competitive adsorption   It is now possible to obtain a highly adsorptive surface in the form of a cloth or fabric. It is produced by reducing the polymer fibres in a woven cloth down to 'pure' carbon whilst retaining the integrity of the fabric. When dipped into a colloidal sol the large surface area and high adsorptivity of the carbon make possible the removal of large quantities of impurity materials from the sol and the surrounding liquid. Success depends upon pretreating the cloth to ensure that it does not release contaminants to the colloid.

Large porous particles could also be used in principle for the same purpose but in practice it is difficult to obtain a material with a high surface area in which the pores are still readily accessible to the colloid and from which it can readily be separated after treatment.

References

Adamson, A. W. (1990). *Physical chemistry of surfaces*, (5th edn). Wiley, New York.

Davies, J. T. and Rideal, E. K. (1963). *Interfacial phenomena*, (2nd edn), p. 360. Academic, New York.

Homola, A. and James, R. O. (1977). *Journal of Colloid and Interface Science*, 59, 123.

See Matijevic, E. and Bell, A. (1973), in *Particle growth in suspension*, (ed. A. L. Smith), pp. 179-93, Academic London, for discussion of the theory, and a long series of papers by Matijevic and his co-workers in *Journal of Colloid and Interface Science* throughout the 1970s and 1980s for details of the procedures for obtaining different sols.

van den Esker, M. W. J. and Peiper, J. H. A. (1975). In *Physical chemistry: enriching topics*, IUPAC Commission 1.6 (ed. H. van Olphen and K.J. Mysels), p. 223. Theorex, La Jolla, California.

Zaiser, E. M. and La Mer, V. K. (1948). *Journal of Colloid and Interface Science* 3, 571.

MICROSCOPIC COLLOIDAL BEHAVIOUR

Perhaps the most important single characteristic of a colloid is the size of the particles of the disperse phase. This is particularly so in the case of solids dispersed in a liquid, which will be our primary concern. Naturally occurring, and most synthetic, colloids are rarely of uniform particle size (an exception would be a suspension of bacterial cells or a pure protein preparation). It is, therefore, necessary to obtain a measure of both the average particle size and the distribution of sizes, that is, the relative numbers of particles in each size range. The methods used to determine particle size rely on a number of fundamental properties of colloids and we will begin by examining them. They happen also to be important properties in their own right: Einstein's theory and Perrin's experiments on Brownian motion of colloidal particles, at the beginning of this century, put the atomic–molecular theory on a sound basis with the first realistic estimates of the Avogadro number.

2.1 Brownian motion and diffusion

Brownian motion refers to the ceaseless jiggling motion which small colloidal particles undergo. It was first studied by an English botanist named Robert Brown in the nineteenth century when he examined with a microscope a suspension of pollen grains in water. He was able to show that it was not confined to living particles and, with the development of the kinetic theory of matter in the 1870s, Brownian motion became recognized as a visible manifestation of the kinetic energy of the solvent molecules. It gave us the first opportunity to 'see' the motions of the molecules, since the collisions with the solvent must quickly bring the particles to have a kinetic energy equal to the average value for the surrounding molecules. Note also that small particles must move much more rapidly than larger ones to carry the same energy.

The same ceaseless motion is also responsible for the process of diffusion, whereby a solute or a gas ultimately becomes uniformly distributed throughout a region to which it has access. Systems containing fairly large (around 1μm) colloidal particles allow us to study these processes by direct microscopic observation and the fact that they depend on particle size can be used to measure that size in any system.

2.1.1 *Distribution of particles in a gravitational field*

Consider, for example, the effect of the earth's gravitational field on a colloidal suspension. If the particle size and density are large enough, equilibrium may only be reached when all of the particles have settled to the bottom of the container. If, however, the particles are small and/or their density is near to that of the solvent, we may get an equilibrium set up in which the number of particles varies with height, just as occurs with gas molecules in the earth's atmosphere. The equilibrium distribution can be calculated in a number of ways; the following thermodynamic argument allows us to introduce an idea which will prove useful in future discussions.

In the absence of any external forces, the composition of a phase at equilibrium is uniform throughout and is characterized by a unique value of the chemical potential (or partial molal Gibbs free energy), μ_i:

$$\mu_i = \bar{G}_i = \left(\frac{\partial G}{\partial n_i}\right)_{p, T, n_j} = \text{a constant} \qquad (2.1)$$

that is,

$$\mathrm{d}\mu_i = 0$$

for each of the species i present. (G is the total Gibbs free energy of the system and n_i is the number of moles of component i.) If an external field is involved, it must be incorporated into eqn (2.1). To consider the effect of the earth's gravitational field, we can introduce the gravitochemical potential $\bar{\mu}_i$ defined by:

$$\bar{\mu}_i = \mu_i + M_i\phi$$

and

$$d\bar{\mu}_i = d\mu_i + M_i d\phi = 0 \qquad (2.2)$$

at equilibrium. Here ϕ is the gravitational potential ($=gh$) and M_i is the molar mass. Then since, in dilute solution,

$$\mu_i \approx \mu_i^0 + RT \ln c_i \qquad (2.3)$$

We could rearrange eqn (2.2) to determine the concentration c_i as a function of height:

$$d \ln c_i = \frac{-M_i g \, dh}{RT}. \qquad (2.4)$$

In colloidal dispersions, gravitational forces are much more significant than they are for molecular solutions or gases, so that the distance scale on which significant concentration differences occur is very much shorter. Thus, in the atmosphere, we would need to take measurements of the gas composition over heights of hundreds of metres to observe significant concentration differences; in a colloidal suspension the same effects can be seen on the surface of a microscope slide.

Perrin and his collaborators in 1909 were able to obtain, by careful cen-

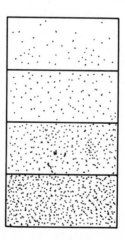

Fig. 2.1. An example of the sedimentation equilibrium of gamboge spheres observed by Perrin (particle radius $= 0.29 \, \mu m$ at levels $10 \, \mu m$ apart).

trifugation, small quantities of uniform spherical particles of a naturally occurring gum called gamboge. When a few drops of this suspension were mounted on a vertical microscope slide and allowed to equilibrate, the number of particles at different heights could be counted directly (Fig. 2.1). If these were substituted in the appropriate integrated form of eqn (2.4):

$$N_i = N_i^0 \exp\left(\frac{-m_i g (h - h_0)}{kT}\right) \tag{2.5}$$

they enabled an estimate to be made of the gas constant, k, per particle. Since $k = R/N_A$, where N_A is the Avogadro number, it was then possible to estimate this last quantity from the well known value of the molar gas constant, R. Perrin obtained a value of $(6-7) \times 10^{23}\,\mathrm{mol}^{-1}$ and subsequent measurements using a more suitable colloid gave 6.05×10^{23}, very close to the now accepted value of 6.022×10^{23}. (Notice that eqn (2.5) could be written down using the Boltzmann expression for the number of particles with different potential energies, $m_i gh$.)

Apart from the problem of obtaining particles of identical size, Perrin had to estimate the mass of the individual particles, which he did by measuring the density of the gum and estimating the size of the particles (which were just a bit too small to see properly in the microscope because of diffraction effects) by an ingenious trick. He was able to get the particles to stick together in strings and to measure the length of a countable number of particles. (Just one of the tricks that earned him the Nobel prize!)

2.1.2 The one-dimensional random walk

The other aspect of Perrin's work which earned him the Nobel prize was his careful study of the random (Brownian) motion of colloidal particles. To estimate how far a particle will move in a given time requires some elementary ideas of probability theory. This material is covered in the standard texts in physical chemistry and we will not reproduce the details here (see Atkins 1982, pp. 908–20). The argument examines how a particle, moving in steps of length l, randomly to left or right along a line, can gradually drift away from its initial position. (The one-dimensional case is easily treated and the generalization to three dimensions is straightforward.) If a step is taken every τ seconds then after time, t, and t/τ steps, the probability that the particle has moved a distance, x, from the starting point is given by:

$$P(x, t) = \left(\frac{2\tau}{\pi t}\right)^{1/2} \exp\left(\frac{-x^2 \tau}{2tl^2}\right). \tag{2.6}$$

Equation (2.6) bears a striking resemblance to the **normal distribution function**, otherwise known as Gauss's error function, or just the Gaussian

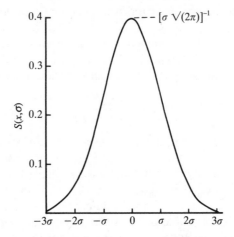

Fig. 2.2. The normal distribution curve. Note that although, theoretically, the distribution stretches to infinity in both directions, in practice it has become negligible after ±3 standard deviations from the mean. The quantity $[\sigma\sqrt{(2\pi)}]^{-1}$ is called the **precision** of the distribution.

(Fig. 2.2) which, appears in numerous contexts to do with probability and statistics. It takes the form

$$P(x, \sigma) = \frac{1}{\sigma\sqrt{2\pi}} \exp\left[-\frac{1}{2}\left(\frac{x - \bar{x}}{\sigma}\right)^2\right] \tag{2.7}$$

where $\bar{x}$ is the mean value of x and σ is called the **standard deviation** of the distribution.

Note that in eqn (2.6) $P(x, t) = P(-x, t)$ but arrival at $-x$ is treated as a different result from arrival at x although they are equiprobable. The standard deviation can be identified with $(t/\tau)^{1/2} l$.

2.1.3 Diffusion

In the absence of any external fields, a phase at equilibrium will have each of its components uniformly distributed throughout. That is expressed thermodynamically by saying that the chemical potential of each component is constant (eqn (2.1)) throughout the phase. Departures from equilibrium occur if one or more components varies in chemical potential (that is activity or concentration) from one point to another in the phase. Whenever that occurs, there is a tendency for the substance to **diffuse** from regions of high concentration to lower concentration until those chemical potential differences are smoothed out. The driving force, f_d, for this diffusion process is the concentration gradient, or more exactly, the (spatial) chemical potential gradient:

$$f_d = -d\mu/dx.$$

(Note that there is no actual force on the particle. It moves randomly but the **net** motion is in such a direction as to maximize the entropy of the system and that will ultimately produce a uniform distribution. This 'phantom' force is merely a convenient way of thinking about the process.) Substituting from eqn (2.3) for the case of a single particle (so that R is replaced by k) we have

$$f_d = -\frac{d}{dx}(\mu_i^0 + kT\ln c_i) = -\frac{kT}{c_i}\frac{dc_i}{dx} \qquad (2.8)$$

since the standard state value is constant throughout the phase. This force sets the particle in motion and, when that happens, it experiences a viscous 'drag' force due to the surrounding fluid. The viscous force, f_v, is proportional to the velocity, u, of the particle:

$$f_v = Bu \qquad (2.9)$$

where B is called the friction coefficient. As the particle increases speed, the viscous force increases until it can balance the diffusion force, and the particle then acquires its terminal diffusing velocity, u_d, so that $f_d = Bu_d$.

The **flux**, J_i, of material is the amount of substance i which passes through a unit of area placed at right angles to the direction of diffusion. It is related to the diffusion velocity by the expression

$$J_i = u_d c \qquad (2.10)$$

as can be seen from Fig. 2.3.

The flux of material is also given by **Fick's first law of diffusion:**

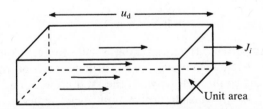

Fig. 2.3. Substance i is diffusing from left to right. The flux of material (mol s^{-1} per unit area) moving across the right-hand face is equal to all the material in the box since the material initially at the far left-hand end can just reach the right-side plane in one second. The volume of the box is u_d in appropriate units and so the flux is $u_d c$.

$$J_i = -\mathfrak{D}\,dc/dx \tag{2.11}$$

where $\mathfrak{D}$ is the **diffusion coefficient**. We can therefore write

$$\mathfrak{D} = -\frac{J_i}{dc/dx} = \frac{-u_d c}{dc/dx} = \frac{-f_d c}{B(dc/dx)}$$

that is,

$$\mathfrak{D} = kT/B \tag{2.12}$$

using eqn (2.8). This is one of many important relations attributed to Einstein.

The friction factor B, which measures how strongly the surrounding fluid resists the motion of the particle, depends on the shape and size of the particle and the viscosity of the fluid. For the simplest case of a smooth spherical particle of radius r immersed in a fluid of viscosity η it is given by the Stokes relation

$$B = 6\pi\eta r. \tag{2.13}$$

For simple geometric shapes like spheroids and cylinders, the calculation of B is more difficult and for irregular bodies it becomes impossible but Atkins (1982, p. 783) gives formulae and some calculations for some of the more common shapes. B can, however, be measured for any system and the result used to obtain information about the particles or macromolecules.

The general equations of diffusion are treated in standard text books (for example, Atkins 1982) where a number of important relations are derived. Fick's first law (eqn (2.11)) can apply only to situations where time is unimportant, that is to steady-state concentrations. That is far too restrictive for most purposes but fortunately one can introduce time effects using **Fick's second law of diffusion**, which is derived by applying the first law and the notion of conservation of mass (Atkins 1982). The result is

$$\frac{\partial c}{\partial t} = \mathfrak{D}\,\frac{\partial^2 c}{\partial x^2}. \tag{2.14}$$

Solving such a partial differential equation calls for techniques which need not concern us. There is, however, one instructive solution which will shed some light on the connection between the random walk and the diffusion process (Atkins 1982). Consider a situation where we begin ($t = 0$) with a thin layer of substance i spread uniformly on one surface of a container (the yz plane), and allow this material to diffuse outwards in the x direction. At time zero the concentration at the surface is c_0 and at subsequent times the total amount of material must remain constant. At each instant in time, the concentration will be constant at all points on a plane parallel to the original source, so c is a function of x and t only. With these conditions the solution of eqn (2.14) turns out to be

$$c(x, t) = c_0 \left(\frac{1}{4\pi\mathfrak{D}t} \right)^{1/2} \exp\left(\frac{-x^2}{4\mathfrak{D}t} \right). \tag{2.15}$$

That result is shown in Fig. 2.4. The similarities between this and eqn (2.6) are obvious, and we can readily make the identification

$$\tau = l^2/2\mathfrak{D}. \tag{2.16}$$

The measure of how far the substance has spread in time t is the **root mean square (r.m.s.) displacement** $\langle x^2 \rangle^{1/2}$. (We use this rather than the mean displacement $\langle x \rangle$ because, for a substance diffusing in both directions, the mean displacement would be zero no matter how far it had spread and that would not be very informative). The r.m.s. displacement is analogous to the standard deviation in the normal distribution curve and comparing eqns (2.6), (2.7), and (2.16) we can write

$$\langle x^2 \rangle^{1/2} = (2\mathfrak{D}t)^{1/2}. \tag{2.17}$$

This is the Einstein–Smoluchowski equation which was used by Perrin in the second procedure for evaluating the Avogadro number. He measured, by

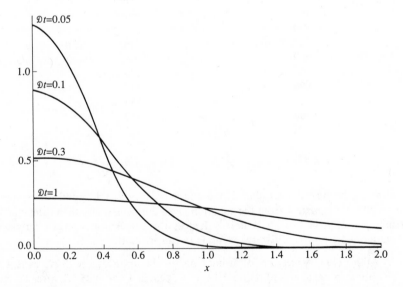

Fig. 2.4. Diffusion of a solute from a plane surface. The ordinate is $c(x,t)/c_0$ for unit area of surface. (From Atkins (1982, p. 909) — note that his ordinate must be divided by 2.)

direct microscopic observation, the r.m.s. displacement of particles after known amounts of time and used them to estimate $\mathfrak{D}$ from eqn (2.17). Then from the relation (eqns (2.12) and (2.13))

$$\mathfrak{D} = kT/B = kT/6\pi\eta r \qquad (2.18)$$

he could again estimate k and, hence, N_A from the gas constant, R (Exercise 2.1.2).

Diffusion coefficients for colloidal particles, including proteins, are much smaller (by several orders of magnitude) than those of ions, so the measurement of $\mathfrak{D}$ by direct means was rather difficult to do until recently. Fortunately, the method of photon correlation spectroscopy, which we will examine in the next section, makes the measurement so easy that it provides a routine method for determining the friction factor B and hence an estimate of particle size.

Exercises

2.1.1 Svedberg gives the following table of Westgren's data for the sedimentation equilibrium of a gold sol under gravity:

Height (μm)	Number of particles
0	889
100	692
200	572
300	426
400	357
500	253
600	217
700	185
800	152
900	125
1000	108
1100	78

Assume the particles have radius 21 nm and density 19.3 g cm^{-3} and the temperature is 20 °C. Estimate k from the equation given in the text, and then calculate N_A assuming that $R = 8.31$ J K^{-1} mol^{-1}. Repeat the calculation using a radius of 22 nm and note how sensitive the result is to this variable.

2.1.2 Atkins (1982) gives the following data for the mean square displacement of rubber latex spheres of radius 2.12×10^{-5} cm in water at 25 °C, after selected time intervals:

t (s)	33	60	90	120
$\langle x^2 \rangle$ (μm^2)	88.2	113.5	128	144.

Use the data to estimate the value of the Avogadro number and comment on the result. (Take the viscosity of water as 8.95×10^{-4} Pa s and the gas constant R as 8.31 J K^{-1} mol^{-1}.)

2.1.3 Estimate the diffusion coefficient for a particle of radius 100 nm in water at a temperature of 25 °C. The corresponding value for a halide ion is about 2×10^{-9} m^2 s^{-1}. To what radius does this correspond? Note that despite the many approximations involved, this is a reasonable value for the radius of an ion, though it might be misleading to use it to infer anything about ion hydration.

2.2 Light scattering

The scattering of light by colloidal particles is one of their more striking characteristics. The light scattered from dust particles in a projector beam, a searchlight or a car's headlights is an example of the **Tyndall effect**. When a light beam is passed through a pure, dust-free molecular solution its path is not visible when viewed from the side. A similar experiment conducted with a colloidal suspension, even if it looks perfectly homogeneous, will reveal the light path because light will be scattered in all directions by the particles. The scattering pattern (that is the intensity of scattered light as a function of the direction θ between the incident and the scattered beam) depends very strongly on the particle size and on the wavelength of the light. The spectral colours that are sometimes generated (as in the case of the rainbow) have fascinated investigators for hundreds of years. The colours of a butterfly's wing, an opal, or mother-of-pearl shell arise from the same source and good potters have been able to reproduce some of these effects in iridescent glazes, using colloidal materials. It is only recently, however, with the advent of lasers capable of giving an intense and narrow beam of monochromatic (that is constant wavelength) light, together with sensitive and stable photon detectors, that the full potential of light scattering as a tool to study macromolecular and colloidal systems has been realized. Rapid data accumulation and analysis by computer have also been important factors.

The basic principles of light scattering are treated in most physical chemistry texts (Atkins 1982). When electromagnetic radiation strikes a particle it may be absorbed, transmitted, scattered, refracted, or diffracted. Absorption occurs when the light is able to excite the material to higher quantum levels and we will assume that it is unimportant in the present discussion. Simple refraction refers to the bending of the light ray as it passes into the particle, if it strikes at an angle; such behaviour is relevant only if the particle is very large compared to the wavelength of the light. For colloidal systems the particle is smaller than or comparable in size to the wavelength and for that reason it is scattering which is the most important process. It is what makes milk and the clouds look white and the sky look blue.

2.2.1 Scattering intensity

A proper treatment of the interaction between light and a particle consisting
of many atoms would require the application of quantum field theory and
would be impossible by present methods. Fortunately, however, it is possible
to tackle the problem using classical methods, and a quite sophisticated solution had been developed to cover all possible wavelengths, even before the
quantum theory had become accepted. In fact, an approximate analysis, for
the case where the particle diameter is much smaller than the wavelength
($d < \lambda/20$) and for refractive indices, n, very close to unity had been given
by Lord Rayleigh in 1871. His equation for the scattered intensity, I, from
a single particle is

$$\frac{I}{I_{0,u}} = \frac{8\pi^4}{\lambda^4 r^2}\left(\frac{\alpha}{4\pi\varepsilon_0}\right)^2 (1 + \cos^2\theta) \qquad (2.19)$$

where $I_{0,u}$ is the initial intensity of **unpolarized** light of wavelength λ, in the
surrounding medium, and α is the particle polarizability. The angle θ is
measured in the plane of the incident and scattered beam and I is the intensity
at a distance r from the particle.

We will not try to derive this equation, but we can see how it arises in the
following way: in the classical theory of scattering one assumes that the electrons in a material in its undisturbed state are in some sort of 'equilibrium'
configuration, that is electrically neutral. When a light wave strikes such a
material, the electric vector of the wave causes the electron to depart from
its normal position and this induces a dipole moment, the magnitude of
which depends on the polarizability of the material. (The polarizability, α,
measures the ease with which the electrons can be displaced.) It can be shown
(Exercise 2.2.1) that the polarizability of a sphere of refractive index n_1 and
radius a is proportional to its volume:

$$\alpha = 4\pi\varepsilon_0 a^3 (n^2 - 1)/(n^2 + 2) \qquad (2.20)$$

where n is the refractive index relative to that of the surrounding medium
($n = n_1/n_0$).

Since the particle is small compared to the wavelength of the light, we can
assume that all parts of the particle are subjected to the same electric field
strength at the same time. The electric vector of the light wave is fluctuating
with frequency, ν say, and so the induced dipole moment, μ, will fluctuate
with the same frequency, though not necessarily in phase with the light wave.
Such a fluctuating dipole must, according to Maxwell's theory, emit electromagnetic energy in all directions at this same frequency, ν. This is the
scattered light, the intensity of which falls off as r^{-2} with distance from the
dipole. The intensity of the radiation in any direction depends on the square

of the electric field, E, and the field strength is proportional to the magnitude of the dipole which, in turn, is proportional to the polarizability, so:

$$I \propto E^2 \propto \mu^2 \propto \alpha^2 \propto a^6$$

(from eqn (2.20). It is then obvious from eqn (2.19) that, to keep the dimensions correct, the intensity must depend on the inverse fourth power of the wavelength. Substituting eqn (2.20) in (2.19) gives, for the total scattering at angle θ, from N_p particles per unit volume:

$$\frac{I_\theta}{I_{0,u}} = \frac{9\pi^2(n_1^2 - n_0^2)^2}{2\lambda^4 r^2(n_1^2 + 2n_0^2)^2} \, v^2 N_p (1 + \cos^2\theta) \qquad (2.21)$$

where v is the particle volume. In this equation the terms 1 and $\cos^2\theta$ refer, respectively, to the vertically and horizontally polarized components of the scattered light (Fig. 2.5). Note that when viewed at 90° only vertically polarised light is visible.

Rayleigh's equation (2.21) is particularly applicable to the scattering from molecules and it is used to explain why the sky is blue: scattering from molecules in the air (or more precisely from fluctuations in the density of molecules in the air) is strongest for light of short wavelength, so it is the blue end of the spectrum which we see when we look in directions away from the sun. At sunset it is the red transmitted light which is most obvious and the scattering is increased by the presence of dust particles in the lower atmosphere.

Rayleigh's equation does not apply to most colloids because the particles are too big, but eqn (2.21) is important because it emphasizes the strong

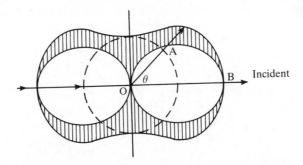

Fig. 2.5. Polar diagram of the light scattered in different directions from a particle at O. The intensity in any direction is measured by the length of the line drawn from the origin to the curve in that direction. The inner (clear) region represents unpolarized light whilst the cross-hatched region represents the (vertically) polarised light. The ratio OA/OB is the depolarization ratio ($I_H/I_V = cos^2\theta$) which varies from 1 in the forward direction to 0 at right angles to the original direction.

dependence of scattering on particle size, on wavelength, and on scattering angle. Since $I_\theta \propto (v^2 N_p)/\lambda^4$ at any particular angle it is apparent that for any particle size (v constant) the scattering increases directly with the particle concentration and at a given mass concentration (vN_p constant) of the particles, the scattering will increase with particle size. This latter property is used to follow the progress of particle aggregation (called **coagulation** or **flocculation**), which is discussed in the next section.

When the particle size is comparable to the wavelength of the light, the scattering pattern becomes much more complicated, because the electrical field strength is no longer the same at all points in the particle at each instant in time. The scattered waves from different parts of the particle interfere with one another and the angular pattern is no longer like that in Fig. 2.5 but many resemble that in Fig. 2.6. Note that, apart from the complexity of the pattern, there is a pronounced emphasis on forward scattering for these larger particles.

2.2.2 *Dynamic light scattering (photon correlation spectroscopy)*

This procedure also goes by the name of quasi-elastic scattering (QELS) which refers to the fact that when a light photon hits a moving particle, the frequency of the scattered light, as measured by a stationary observer, will be different from the incident frequency. The frequency is slightly increased or decreased depending on whether the particle is moving towards or away from the observer. This is called **Doppler broadening** (Fig. 2.7) and from the extent of the broadening we can determine the diffusion coefficient.

The intensity of the scattered light of frequency ω can be represented by the function

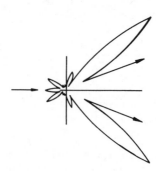

Fig. 2.6. Scattering pattern of light from a spherical particle for which $2\pi a/\lambda = 6$ and $n_1/n_0 = 1.44$; calculated from the theory due to Mie. (After Kruyt (1952).)

$$I(\omega) = A_1 \frac{\mathfrak{D}Q^2}{(\omega - \omega_0)^2 + (\mathfrak{D}Q^2)^2} \tag{2.22}$$

where ω_0 is the central (that is, incident) frequency, A_1 is a constant and Q is the magnitude of the scattering vector:

$$Q = \frac{4\pi n_0}{\lambda} \sin(\theta/2) \tag{2.23}$$

where n_0 is the refractive index of the medium and θ is the scattering angle. Q measures how strongly the light interacts with the particles. Small Q values correspond to glancing incidence of the photon with the particle, whilst for θ approaching 2π the collision is essentially head-on.

The width of the peak at half-height ($\Delta\omega_{1/2}$) is directly related to the diffusion coefficient (Exercise 2.2.2):

$$\Delta\omega_{1/2} = \mathfrak{D}Q^2. \tag{2.24}$$

The function shown in Fig. 2.7 is called a Lorentzian (as distinct from Gaussian (Fig. 2.2)) distribution and it is frequently encountered in the discussion of light absorption.

The use of laser sources has made it comparatively easy to measure these tiny changes in frequency and hence to exploit the consequences of Eqn (2.22). The initial experiments in the field were done by mixing the scattered light with the incident beam so as to create a **beat frequency** ($\omega - \omega_0$); a spectrum analyser was then used to determine the intensity of that signal for different ω values (Fig. 2.7) and so determine the diffusion coefficient from eqn (2.24). The mean radius could then be calculated using eqn (2.18). Modern instruments use a different procedure called **photon correlation** (§3.5.2) to extract the same information.

Exercise

2.2.1 Atkins (1982) shows that for a system of N non-polar molecules per unit volume of refractive index n the polarisability, $\alpha N = 3\varepsilon_0(n^2 - 1)/(n^2 + 2)$. Show that this is consistent with eqn (2.20).

2.2.2 Establish eqn (2.24) from (2.22).

2.3 Coagulation and flocculation — stability and instability

So far we have treated colloidal systems as consisting of a large number of small particles freely moving about in a suspending fluid. We must recognize, however, that these particles will be colliding with one another very frequently, in the course of their Brownian motion, and in a collision they may be so attracted to one another that they remain stuck together. This

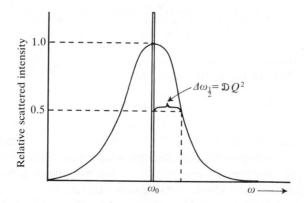

Fig. 2.7. Doppler line broadening. The scattered intensity at a particular angle for a stationary particle is given by the central sharp line. If the particle is moving, the line becomes (frequency) broadened and the intensity at the central frequency (ω_0) is reduced. In colloidal systems the broadening is a direct reflection of the random motion of the particles; it amounts to about 1000 Hz in a background of $\omega_0 \sim 10^{14}$ Hz.

doublet will then move more slowly than the single particles but will still encounter other particles which may stick so that the aggregate grows to become very large and much less mobile, until it is so large that it settles rapidly from the suspension. We noted in §1.3 that we are here most concerned with **lyophobic** or irreversible colloids for which the most stable thermodynamic state is one in which the particles are all condensed together into a large lump. Such systems can remain as individual particles only if there is some mechanism to prevent them from sticking together when they collide with one another. The system is then said to be **stable** in the colloid chemical sense. What is it that prevents that aggregation process?

There are at least two ways to produce stability:

(1) the particles can be given an electric charge (either positive or negative) and if all have the same charge then they will repel one another on close approach; or

(2) the surface of the particles can be coated with an adsorbed layer of some substance (say a polymer) which itself can prevent close approach.

Method (1) is referred to as **electrostatic stabilization** and method (2) is called **steric stabilization**. The first name is unobjectionable but the second name is perhaps a little misleading. It suggests that the adsorbed material is simply filling the space around the particle and acting as a steric barrier to approach. In the case of an adsorbed polymer layer the situation is more subtle. The polymer is normally interacting strongly with the suspending medium and is not packed tightly against the surface. Rather it is like a fuzzy or springy coating which cannot be easily compressed because compression forces the polymer segments together and restricts their interaction with the surrounding liquid. That would lead to an increase in free energy, so it shows up as a repulsive force. In some respects the polymer is able to convert the lyophobic colloid into a lyophilic one.

A system is said to be colloidally **unstable** if collisions lead to the formation of aggregates: such a process is called **coagulation** or **flocculation**. It is a very striking phenomenon because when the system changes from being stable to unstable its whole character and most of its properties (like settling, filtration, and flow behaviour) change in a profound way. Much attention is therefore paid to the control of colloid stability. We will discuss this in more detail in Chapter 9. For the moment we need only look at the broad features.

2.3.1 The electrical charge at a surface

We saw in §1.4.4 that clay minerals carried a surface charge due to defects in the crystal structure. These charges were balanced by an equal number of oppositely charged ions adsorbed on the basal surfaces of the particles. Most colloidal particles are electrically charged, as can be readily demonstrated by watching their motion in an electric field. The origin of the charge does not, however, always lie in the defect structure of their crystals. Some particles have surface groups (like–COOH,–OSO$_3$H, and NH$_2$) which can react with either acid or base to provide stabilizing charges. Most metal oxides, which are an important class of colloids, have a surface layer of the metal hydroxide which is amphoteric in nature and can become either positively or negatively charged depending on the pH:

$$M\text{–}OH + OH^- \rightarrow M\text{–}O^- + H_2O$$
$$M\text{–}OH + H^+ \rightarrow MOH_2{}^+. \tag{2.25}$$

When solid particles are immersed in a fluid there is a general tendency for ions of one sign to be preferentially adsorbed onto the solid and for the oppositely charged ions to remain in the neighbouring fluid. In some cases the origin of the ions is obvious. When, for example, silver iodide particles are suspended in water, they must establish equilibrium according to their solubility product, K_s, so that the activities of silver and iodide ions in solution satisfy:

$$a_{Ag^+} a_{I^-} = K_s. \tag{2.26}$$

The activity, or concentration of either the silver or the iodide ion can be altered by adding some silver nitrate or some potassium iodide. The concentration of the other ion will then adjust itself to satisfy eqn (2.26) by forming an appropriate amount of the solid. We will find (§7.3.1) that I^- ions are adsorbed much more readily from solution onto the AgI crystal than are Ag^+ ions. The net charge and, hence, the **electrostatic potential** on the crystal surface, relative to the surrounding liquid, is strongly dependent on the balance between I^- and Ag^+ ions in the solution. The ions Ag^+ and I^- are therefore called **potential-determining ions** for the silver iodide system. In the same way, the ions H^+ and OH^- are potential-determining ions for the oxide systems, along with a great many other colloids like the polymer latexes and many biological surfaces. In such systems the surface charge and potential are determined largely by the balance between H^+ and OH^- in the solution (that is by the pH).

 Figure 2.8 shows schematically what the electrostatic potential in the neighbourhood of a positively charged colloidal particle is expected to look like. It is similar to that near an ion in an electrolyte solution. Negative counterions (§1.4.2) are attracted towards the particle by the electric field generated by the positively charged surface but they are also subject to thermal motion, which tends to spread them uniformly through the surrounding medium. The result is a compromise in which a few negative ions remain very close to the surface and their concentration gradually falls off away from the surface, until it reaches that of the bulk solution. The distance over which this occurs depends upon the electrolyte concentration. It may be only a fraction of a nanometre at high concentrations (approaching 1 M), but in very dilute solution (say, 10^{-5} M) the region of varying concentration (the 'space charge' as it is sometimes called) can stretch out for a few hundred

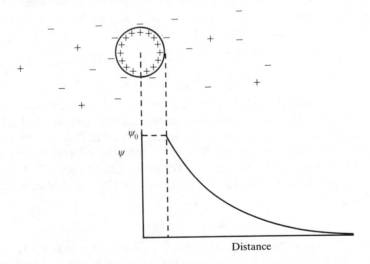

Fig. 2.8. Electrostatic potential near a positively charged colloidal particle.

nanometres. This charge arrangements is called the **diffuse electrical double layer** around the particle. Adding salt to a colloidal suspension causes the diffuse double layer to shrink around the particles, a process known as **double-layer compression**; we will find that this has important consequences for the stability of the colloid.

2.3.2 Observation of coagulation behaviour

One can study the coagulation process at the microscopic or the macroscopic level. At the microscopic level, we would study the frequency of collisions between individual particles and the rate of formation of doublets, triplets, etc. and hope to relate this to the (diffusive) motion of the particles and the forces between them. We noted in §2.2 that light scattering can be used to follow these early stages in the coagulation process and we will discuss that in more detail in Chapter 9.

In a lyophobic colloid some such coagulation process is always going on, even at low electrolyte concentration when the sol may appear, macroscopically, to be quite stable. This is to be expected if, as suggested in §1.3, these sols are thermodynamically unstable with respect to aggregation and are rendered 'stable' only in a kinetic sense. The **rate** of coagulation can be extremely slow. In this regime the presence of a **stabilizer** (either an electrical charge or an adsorbed layer) sets up a repulsion barrier between the approaching particles. This decreases the efficiency of collisions so that perhaps only 1 in 10^5 or 1 in 10^{10} collisions may result in particle–particle contact and a permanent doublet.

If coagulation continues for long enough the aggregates become large enough to be macroscopically visible. They are then called **flocs** and if they differ in density from the surrounding fluid they will settle quite rapidly, leaving a more or less clear **supernatant** liquid. If the floc density is lower than that of the liquid, the aggregates move upwards in a process called **creaming** for obvious reasons.

2.3.3 Coagulation by potential control

One method of determining iodide ion concentration in solutions is to titrate the unknown with a solution of silver nitrate of known concentration and to follow the precipitation of the silver iodide. The end-point can be detected by adding a few drops of potassium chromate, which does not enter into reaction until all the iodide is precipitated. What concerns us here is the appearance of the precipitate in the neighbourhood of the end-point.

Silver iodide is a very insoluble salt (with a solubility product of about 8×10^{-17}) and when Ag^+ and I^- ions are mixed in solution the precipitation occurs rapidly and the particle size is very small — frequently of colloidal dimensions. The smooth milky appearance of the precipitate is characteristic of a stable colloidal dispersion. The sol is formed in the presence of an excess of iodide ions and so is negatively charged (§2.3.1). As more silver ions are added, the excess iodide ions gradually disappear and silver ions can compete more effectively for the surface sites. The negative charge on the surface decreases to such an extent that the repulsive force between the particles disappears. The sol then becomes unstable and the particles begin to coagulate; large flocs of silver iodide appear in the titration vessel just as the reddish colour of the precipitating silver chromate heralds the occurrence of the end-point.

This is a clear case of coagulation by **potential control**. The electric charge on the particles at some point (near the end-point) is zero and this is called the **point of zero charge** (p.z.c.) for the colloid. It is found to occur at a particular activity of the Ag^+ ions, namely at a concentration of $3.2 \times 10^{-6} M$, or $pAg = -\log_{10}[Ag^+]$ of 5.5. This corresponds to an iodide ion concentration of $8 \times 10^{-17}/3.2 \times 10^{-6} = 2.5 \times 10^{-11} M$, indeed very near to the end-point.

2.3.4 Coagulation by electrolyte addition

The size of the repulsion barrier which stabilizes a colloidal suspension depends on the nature of the surface. For an electrostatically stabilized sol, the barrier depends on the magnitude of the charge on the surface and the extent of the diffuse double layer around the particles, which in turn depends on the total electrolyte concentration as noted in §2.3.1. It is necessary to distinguish between the role of the potential determining ions (p.d.i.) and the other ions (called **indifferent ions**) which have no special relationship with the surface. If the p.d.i. concentration is adjusted so that there is a large excess of either positive or negative charges at the surface, the stability of the colloid might be expected to be high, but it can still be influenced by changing the extent of the diffuse double layer. If, for example, ions of another salt, like potassium nitrate, are added to the sol this leads to a compression of the diffuse double layer, a smaller repulsion force between approaching particles and ultimately to the phenomenon of **rapid coagulation**. This occurs when the repulsion has been effectively eliminated and every collision results in particle contact. The rate of coagulation is then determined by the rate at which particles can diffuse towards one another due to their Brownian motion.

The electrolyte concentration at which slow coagulation turns into rapid coagulation can be determined fairly accurately and is an important characteristic of the colloid. It is called the **critical coagulation concentration** (c.c.c.). It can be determined by taking a series of test tubes with the same concentration of the colloidal particles. One then adds increasing amounts

Table 2.1 Coagulation concentrations of simple electrolytes (millimole l^{-1}). (From Overbeek in Kruyt (1952, p. 82), with permission.)

Valency	As_2S_3 sol (Negative)		Au sol (Negative)		$Fe(OH)_3$ (Positive)	
Monovalent	LiCl	58			NaCl	9.25
	NaCl	51	NaCl	24	½BaCl$_2$	9.65
	KNO$_3$	50	KNO$_3$	23	KNO$_3$	12
Divalent	MgCl$_2$	0.72	CaCl$_2$	0.41	K$_2$SO$_4$	0.205
	MgSO$_4$	0.81	BaCl$_2$	0.35	MgSO$_4$	0.22
	ZnCl$_2$	0.69			K$_2$Cr$_2$O$_7$	0.195
Trivalent	AlCl$_3$	0.093				
	½Al$_2$(SO$_4$)$_3$	0.096	½Al$_2$(SO$_4$)$_3$	0.009		
	Ce(NO$_3$)$_3$	0.080	Ce(NO$_3$)$_3$	0.003		

of an electrolyte to each tube in turn; the contents are vigorously shaken for a few seconds and then allowed to stand for a couple of hours. The tubes are then again mixed by gently turning them over a few times and allowed to stand for a few minutes. If the salt concentrations have been chosen in the right range, the c.c.c. will appear as that concentration above which the sol settles to leave a clear supernatant. At concentrations below the c.c.c there may be some coagulation but the supernatant remains cloudy. (It is the second mixing which is most important because it allows the flocs formed in the first stage to sweep through the sol and collect up the remaining particles, if there is no barrier to coagulation.)

Even with this quite crude experiment it is possible to see some striking regularities in the behaviour of common colloidal suspensions (Table 2.1).

2.3.5 Critical coagulation concentration

Table 2.1 gives the values of electrolyte concentration required to induce rapid coagulation in three typical inorganic colloidal systems, two negatively charged and one positively charged. Note that:

(1) the coagulation concentration for similar electrolyte solutions is similar but not identical;

(2) the effectiveness of an electrolyte as a coagulating agent generally increases when it contains multivalent ions; and

(3) it is the valency of the *counterion* which is most important in determining the coagulation concentration. Thus for the positively charged $Fe(OH)_3$ sol it is the chloride ion concentration in BaCl$_2$ that determines the coagulation behaviour, not the barium ion concentration.

The strong dependence of coagulation concentration on the valency of the electrolyte is referred to as the Schulze–Hardy rule and has been recognized since the end of the nineteenth century. The theory of stability of lyophobic colloids provides an excellent rationale for this rule. Like all theories it works best for very well characterized and simple systems, but the ideas involved provide an essential framework for the discussion of the complex interactions that occur in real systems, ranging from the swelling of soils, to the structure of the chloroplasts which trap light in the leaves of plants, and from the separation of mineral ores to the treatment of sewerage sludge. The theory is based on the recognition of two forces in any electrostatically stabilized sol: the electrostatic repulsion which opposes aggregation and a universal attractive force, which acts to bind particles together if they can come into close enough contact. It is referred to as the DLVO theory after the four scientists — Deryaguin, Landau, Verwey, and Overbeek — who were largely responsible for its development.

The attractive force, which occurs between any two particles of the same material immersed in a solvent is called the **long-range van der Waals force**. It has its origins in those same forces which act between atoms or molecules to produce the condensation from gas to liquid: the London dispersion interaction (Atkins 1982, p. 783). In the case of colloidal particles, it is assumed that these dispersion forces are more or less additive, so that the interaction between the two particles is estimated by summing the interactions between the atoms of each particle. The result is a force which has a much longer range than the attractive force between individual atoms or molecules. The long-range van der Waals force decreases as the inverse square of the separation and typically extends for distances of the order of the particle size. (Compare this with the force between atoms which extends for only about 1 nm or so.)

2.3.6 Heterocoagulation

In §1.4.4 we discussed some properties of kaolinite particles and, in particular, the origin of the negative charge which occurs on the cleavage faces of those crystals. The structure shown in Fig. 1.9 suggests that the crystal can extend to infinite dimensions in the horizontal direction but in practice the crystals are limited in size to about 1μm. When the sheet breaks it exposes aluminium atoms attached to oxygens and, at low pH, these can accept a proton to generate a positive charge on the edge of the crystal lattice:

$$-Al-O-Al + H^+ \rightarrow -Al-OH^+-Al. \qquad (2.27)$$

We then have a situation where the same crystal can have negative charges on the flat faces and positive charges around the edges. In dilute acid conditions (pH 4–6) the particles can become associated in open card-house struc-

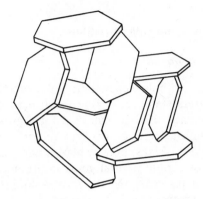

Fig. 2.9. Card-house structure of kaolinite. (After Street (1956).)

tures (like that shown in Fig. 2.9) which gradually open out or collapse as the pH is raised and the positive charges on the edges become neutralized. At high pH the clay can coagulate only if the salt concentration is high enough to reduce the repulsion between the face charges. These ideas have been used to explain the rather subtle changes which occur in properties when the pH is changed. Properties like the flow behaviour are very important in many of the applications of kaolinite, like the making of ceramics and high-quality paper.

There are many other natural systems where the particles are so heterogeneous that individual crystals may have different electrical charges on different faces. Biological surfaces, or those covered with protein, will have both sorts of charge distributed over the surface at some pH values. When such surfaces come together they will tend to align themselves so as to maximize the interactions between positive and negative charges. (This sort of process may be involved in the way biological cells recognize one another when an organism is developing.)

In industrial systems we often encounter mixtures of two or more very different sorts of particles with different surface characteristics. It is then possible for one set of particles to be of opposite sign to the other at a particular pH. The interaction between surfaces having significantly different charge status is called **heterocoagulation**. It is important in processes like surface coating and the preparation of films of one material on another and in mineral processing.

2.4 Effect of polymers on colloid stability

2.4.1 Steric stabilization

We have so far discussed only electrostatic stabilization, and yet the alternative procedure of steric stabilization is probably more widespread, and has certainly been consciously used by mankind for a great deal longer, even though its mechanism of operation has only recently become amenable to study. The early makers of inks, pottery glazes, and paints must have been aware of the value of certain natural gums in promoting stability (in the colloid sense) of their pigments and the early colloid scientists referred to this phenomenon as **protection**.

In effect, the surface of the lyophobic colloid was coated with a layer of lyophilic material of a polymeric (or, at any rate, of a long-chain) nature. When two such particles approach, the interaction between the adsorbed chains causes a repulsion that can be sufficient to induce stability. Its magnitude can be calculated by estimating the effect of the particle separation on the free energy of the adsorbed molecules. As they are forced together the number of chain–chain contacts is increased at the expense of chain–solvent interactions and for a lyophilic (solvent-loving) polymer, that must lead to an increase in the free energy, or a repulsive force (Fig. 2.10). Steric stabilization is important in water, but it is much more important in non-aqueous solvents where electrostatic stabilization is rarely possible.

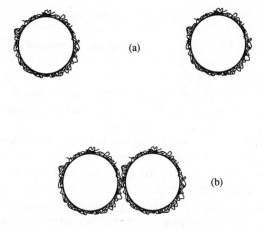

Fig. 2.10. Interaction between two sterically stabilized particles. (a) Separated particles with their adsorbed polymer chains interacting with the solvent. (b) Interpenetration of the polymer chains reduces the amount of polymer/solvent interaction and forces more polymer/polymer contacts, which increases the Gibbs free energy, assuming that the normal conformation in (a) is the equilibrium one.

The free energy of mixing of the adsorbed molecular chains, as the particles come together, can be broken into an enthalpic and an entropic part:

$$\Delta G_{mix} = \Delta H_{mix} - T\Delta S_{mix}. \qquad (2.28)$$

A positive value of ΔG_{mix} (repulsion) can arise either from a positive value of ΔH_{mix} (**enthalpic stabilization**) or a negative value of ΔS_{mix} (**entropic stabilization**) or both. (See Exercise 2.4.1.) It is important to note that these remarks about the mechanism of stabilization apply particularly to systems in which the adsorbed polymer is anchored firmly to the surface so that the chains cannot desorb or move out of the way as the particles approach each other.

2.4.2 Polymer flocculation

Another common method of inducing aggregation of a colloidal dispersion is by the use of a polymeric flocculating agent. (The term **flocculation** is used here for the formation of rather loose aggregates of particles linked together by a polymer, as distinct from **coagulation** where the particles come into rather closer contact as a result of changes in the electrical double layer around the particles.)

A polymer can adsorb on the surface of a colloidal particle as a result of a Coulombic (charge–charge) interaction, dipole–dipole interactions, hydrogen bonding, or van der Waals interactions, or some combination of any or all of these forces. A balance must be struck between the affinity of the polymer groups for the surface, and their interaction with the solvent. The usual result is that the polymer sticks to certain points on the surfaces, but for much of its length it is able to trail out into the solvent. The result is some segments attached to the surface (as **trains**) separated from one another by **loops** which extend into the solution and ending in a (usually quite extended) **tail** at either end of the polymer chain (Fig. 2.11).

When two particles are brought together it becomes possible for the loops and tails of one polymer to attach themselves to bare patches on the

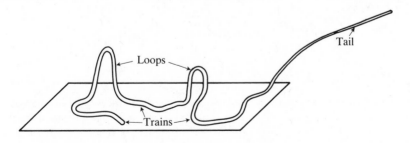

Fig. 2.11. Schematic diagram of polymer adsorbed on a surface.

approaching particle to form **bridges**. This process is assisted if the adsorption density of the polymer is not too high. Indeed, one usually finds that the optimum polymer concentration to achieve flocculation corresponds to about half surface coverage for the polymer. Careful mixing and control of the nature and concentration of the polymer can result in very effective aggregation of the colloid and the method is widely used in industry in the treatment of mineral ores and in purification of waste water.

Exercise

2.4.1 Why is it that enthalpically stabilized systems can be made to coagulate at sufficiently high temperatures, whilst entropically stabilized systems tend to become unstable below some particular temperature?

References

Atkins, P. W. (1982). *Physical chemistry*, (2nd edn), pp. 783, 908–20. Oxford University Press.
Kruyt, H. R. (1952). *Colloid science*, Vol. 1, p. 100. Elsevier, Amsterdam.
Street, N. (1956). *Australian Journal of Chemistry*, **9**, 467.

3

DETERMINATION OF PARTICLE SIZE

3.1 General considerations
3.1.1 Size and shape
3.1.2 Size distributions
    (a) The mean and standard deviation
    (b) Other mean values
    (c) The continuous distribution function
    (d) Theoretical distribution functions

3.2 Microscopy
3.2.1 Optical (light) microscopy
3.2.2 Electron microscopy
    (a) Transmission electron microscope
    (b) Scanning electron microscope
    (c) Atomic force microscope

3.3 Sedimentation methods
3.3.1 Sedimentation under gravity
3.3.2 Centrifugal sedimentation
3.3.3 Instrumentation

3.4 Electrical pulse counting

3.5 Light scattering methods
3.5.1 Intensity methods
    (a) Rayleigh scattering
    (b) Rayleigh–Gans–Debye (RGD) scattering
    (c) Mie scattering
    (d) Fraunhofer diffraction
3.5.2 Dynamic light scattering (QELS or PCS)

3.6 Hydrodynamic methods
3.6.1 Capillary hydrodynamic fractionation
3.6.2 Field flow fractionation (FFF)

3.7 Electroacoustics

3.8 Summary of sizing methods

3.1　General considerations

3.1.1　Size and shape

The size and shape of colloidal particles are amongst their most important characteristics because they determine many other features of the behaviour of colloidal suspensions. The rate of settling, the ease with which they can be filtered, and their flow properties when poured or pumped through a pipe, all depend on particle size. So too does their capacity to adsorb contaminants and, in the case of catalysts, their capacity per unit mass which normally depends on the available surface area, onto which reacting molecules must be adsorbed.

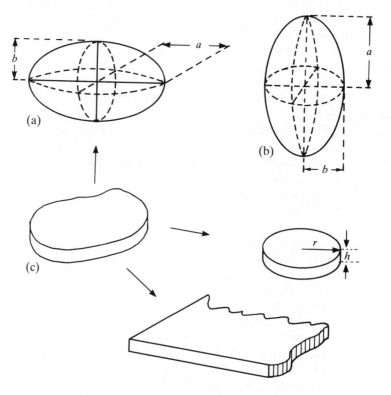

Fig. 3.1. (a) An oblate spheroid, obtained by rotating an ellipse around its short axis. The cross section in the plane where a is measured is circular. (Sometimes described as 'lozenge shaped'.) (b) A prolate spheroid, obtained by rotating an ellipse around its long axis (like a rugby or gridiron football, or a cigar). Sections parallel to b are circular. (c) A disc may be approximated as an oblate spheroid, a cylinder, or a flat plate.

Colloidal suspensions often exhibit a wide range of particle sizes. It is not uncommon, especially for sols produced by grinding (§1.4), for the size of particles to extend over several orders of magnitude. For some purposes it may be enough to know the minimum size, or the maximum, or some sort of average value, but in many cases it is necessary to have a more or less complete knowledge of the distribution, that is the number of particles in each size range.

It must also be recognized that colloidal particles in a sample often vary widely in shape. Unless special care has been taken to produce monodisperse, and/or spherical particles, (§1.4) the particles may be disc or rod shaped or they may be highly irregular, varying from particle to particle and from sample to sample. In all of our theoretical analyses we will be forced to deal only with the simplest shape, a sphere of a certain size. More sophisticated treatments can partially extend the treatment to more complicated shapes like discs and rods provided one can specify the ratio of the major to the minor axis (Fig. 3.1). One practical way of overcoming this limitation, widely used in industry, is to measure the 'size' by a variety of different methods. Since each method interacts with the particles in a different way, one obtains different results from the different methods (except in the case of monodisperse spherical particles). The differences between the different methods can then be used to gather qualitative information about the system. Sometimes it is enough to simply determine which of the available methods gives the 'size' that makes the most sense for one's particular industrial control problem. The important point to note is that different sizing methods can be expected to give different results *because they are measuring different aspects of the size distribution.*

Table 3.1 Specimen size distribution for a precipitate: $\Sigma n_i = 2500 = N$, $\Sigma f_i = 1.00$, $f_i = n_i/\Sigma n_i$

Class range (nm)	Mid-point of class range, d_i(nm)	Number of particles, n_i	Fraction in this class, f_i	Total number with $d < d_i$	Cumulative per cent
51–100	75	29	0.012	29	1.2
101–150	125	109	0.044	138	5.5
151–200	175	211	0.084	349	14.0
201–250	225	372	0.149	721	28.8
251–300	275	558	0.223	1279	51.2
301–350	325	440	0.176	1719	68.8
351–400	375	307	0.123	2026	81.0
401–450	425	223	0.089	2249	90.0
451–500	475	139	0.056	2388	95.5
501–550	525	81	0.032	2469	98.8
551–600	575	31	0.012	2500	100.0

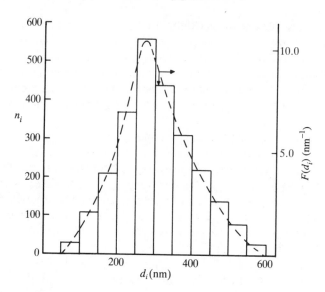

Fig. 3.2. The frequency histogram can be replaced by a smoothed curve. Note that the modal size is the most common one.

3.1.2 Size distributions

Whenever we are confronted with a problem of describing the particle size of a system that is **heterodisperse** or **polydisperse** (that is, contains many different sizes of particles) we resort to breaking the range of sizes up into convenient steps or classes, and recording the number of particles in each class. Consider, for example, the data in Table 3.1, which may represent the diameters of a sample of particles produced by a precipitation reaction. If the observed particle sizes ranged from 65 nm to 0.6 μm one might choose to break the range up into 11 steps of 50 nm, as shown, and to record the number of particles in each class. The resulting data can then be plotted as a histogram or as a smooth curve (Fig. 3.2) or as a curve showing the cumulative percentage equal to or smaller than a given size (Fig. 3.3, called the 'per cent undersize'). Rather than going to the trouble of plotting the data each time, it is often sufficient to specify the main features of the distribution using a few numbers.

(a) The mean and standard deviation   The most important characteristics of a distribution are the **mean,** which measures the central tendency and the **standard deviation** which measures the spread of the data. The mean diameter is defined as

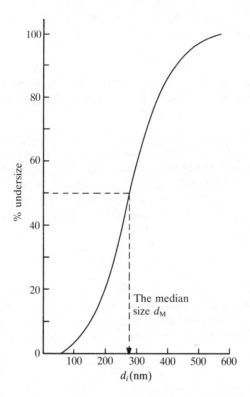

Fig. 3.3. The cumulative frequency curve showing the median size d_M, which divides the distribution evenly (50 per cent of the material has $d_i \le d_M$).

$$\bar{d} = \frac{\Sigma n_i d_i}{\Sigma n_i} = \frac{\Sigma n_i d_i}{N} = \sum f_i d_i \qquad (3.1)$$

where f_i is the fraction in class i (Table 3.1). (To distinguish it from other mean diameters we should strictly refer to this as the number–length mean diameter.)

The standard deviation, σ, is defined as

$$\sigma = \left(\frac{\Sigma n_i (d_i - \bar{d})^2}{N}\right)^{1/2} = \left(\sum f_i (d_i - \bar{d})^2\right)^{1/2}. \qquad (3.2)$$

Note again that we do not use $d_i - \bar{d}$ as the measure of the spread because it can be positive or negative and, for a symmetrical distribution, $\Sigma(d_i - \bar{d}) = 0$ even though the distribution might be quite broad.

The quantity inside the large round brackets is called the **variance**($= \sigma^2$) of the population. We take the square root of the sum of the squares so that the deviation can be more readily compared to the mean. It is not difficult to show (Exercise 3.1.2) that

$$\sigma = (\overline{d^2} - (\bar{d})^2)^{1/2} \tag{3.3}$$

which is often easier to compute than eqn (3.2). (The first term in the brackets is the average value of the squares of the sample diameters.)

(b) Other mean values   If a colloidal suspension is **polydisperse** (that is has many different particle sizes) then the mean value can be expressed in a variety of ways. Different measurement methods will not only make different estimates of the mean, as defined in eqn (3.1), but will also measure different mean values. To see how this comes about, consider the total surface area, A_s, of a polydisperse sample of spherical particles. We can write

$$A_s = \sum n_i \pi d_i^2. \tag{3.4}$$

(This total area can be measured by determining the capacity of the solid to adsorb a gas, provided we know the area occupied by each gas molecule.) If the total number of particles is also known, then the number–area mean diameter, $\bar{d}_{NA}$ is defined as the diameter of the sphere for which

$$\pi \bar{d}_{NA}^2 = \frac{A_s}{N} \tag{3.5}$$

so that

$$\bar{d}_{NA} = \left(\sum f_i d_i^2\right)^{1/2} = (\overline{d^2})^{1/2}. \tag{3.6}$$

Note that a system of N uniform spheres of diameter $\bar{d}_{NA}$ has the same surface area as the original sample, S. In a similar way one can define a series of different average sizes, many of which are directly measurable. One very important one is the volume average particle size which happens to be the one most easily extracted from light scattering methods and is closely related to the mass average size determined from ultrasonic measurements using the ESA technique (§3.7) (see Exercise 3.1.3). It is defined as

$$\bar{d}_{NV} = \left(\frac{\sum n_i V_i}{N}\right)^{1/3} = K\left(\frac{\sum n_i d_i^3}{N}\right)^{1/3}. \tag{3.7}$$

(c) The continuous distribution function   Figure 3.2 shows that the histogram can be replaced by a smooth curve but it is important to note that this is not a plot of n_i against d_i. For the continuous curve, $F(d_i)dd_i$, the number of particles dn_i in the range d_i to $d_i + dd_i$ is given by

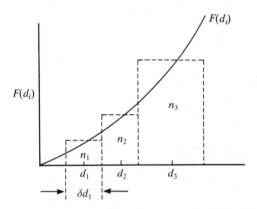

Fig. 3.4. The continuous distribution function, $F(d_i)$, and its relation to the frequency histogram.

$$dn_i = F(d_i)\, dd_i \qquad (3.8)$$

where the function $F(d_i)$ is called the (number) distribution function for d_i. In Fig. 3.2 all of the classes have the same width, so that the height of the rectangles is proportional to n_i. In the more general case we may choose to vary the width of the classes and in that case it is preferable to draw rectangles whose **areas** reflect the numbers of particles in the class. Then we can readily relate the distribution function F to n_i using eqn (3.8). It is apparent from Fig. 3.4 that

$$n_i \approx F(d_i)\delta d_i \qquad (3.9)$$

and the area under the $F(d_i)$ curve will give the total number of particles.

For the data in Table 3.1 the value of $F(d_i)$ is related to f_i by a constant factor, since

$$f_i = \frac{n_i}{N} = \frac{F(d_i)\delta d_i}{N}. \qquad (3.10)$$

One could generate the distribution function (in nm^{-1}), then, by multiplying column 4 by $2500/50 = 50$ and it would have the same shape as the broken curve in Fig. 3.2.

(d) Theoretical distribution functions There are a few 'theoretical' distribution functions which are particularly significant since they closely approximate the distributions found in nature. Some have well founded theoretical bases whereas others are more empirical. The most commonly used and important of the theoretical distribution functions is the **normal** or **Gaussian**

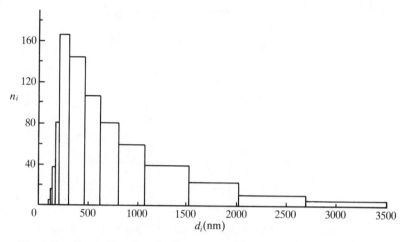

Fig. 3.5. A possible size distribution produced by grinding or crushing.

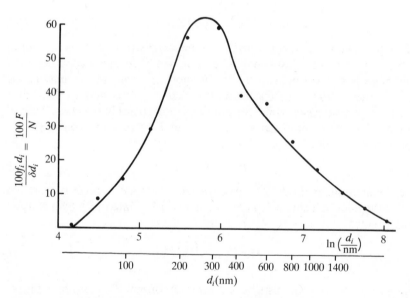

Fig. 3.6. The relative percentage frequency distribution function plotted against $\ln(d_i/\text{nm})$. Data from Fig. 3.5.

distribution which we have already encountered (Fig. 2.2). It is closely related to the error function and lies at the heart of the statistical treatment of errors. It will apply to particle size distributions only in cases where the departure from the mean value is as likely to be positive as negative and where departures become less likely the larger they are. That is not always

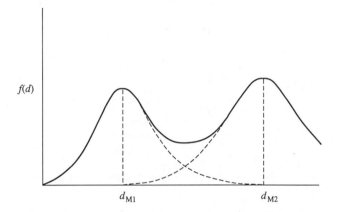

Fig. 3.7. A bimodal distribution resulting from the overlap of two 'normal' distributions having modes d_{M1} and d_{M2}.

the case. In fact, the distribution of sizes obtained from grinding a solid down to form small particles is often very far from 'normal' or Gaussian (Fig. 3.5). In that case it often helps to transform the data by looking at the distribution of log d which may look more like a normal distribution (Fig. 3.6). The advantage of this procedure is that when we are dealing with a normal distribution we can apply all the standard methods of statistics with complete confidence.

Another common complication comes in the form shown in Fig. 3.7 which shows a **bimodal** distribution. Such distributions can occur in the preparation of a sol using the condensation procedure (§1.4.1) if there are two different nucleation periods separated by a significant time interval. The preferred procedure in this case is to separate (using an appropriate mathematical technique called **deconvolution**) the distribution into the underlying populations, assuming that they have some easily represented (for example Gaussian) shape.

Exercises

3.1.1 Calculate the mean, standard deviation, and the variance of the distribution shown in Table 3.1. What is the difference between the mean and the mode in this case?

3.1.2 Establish eqn (3.3) from (3.2).

3.1.3 (a) What would be the value of K in eqn (3.7)?
(b) The (number) mass average diameter size is defined by analogy to the volume average (eqn (3.7)). Set up the appropriate expression and establish the corresponding value of K in terms of the density of the solid.

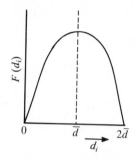

Fig. 3.8. Parabolic distribution (Exercise 3.1.5).

3.1.4 Calculate the number–area mean diameter, $\bar{d}_{NA}$, of the particles described in Table 3.1 and compare it with the number–length mean diameter. Show that $\bar{d}_{NA} = (\sigma^2 + \bar{d}^2)^{1/2}$ and check this with the result obtained in Exercise 3.1.1.

3.1.5 The accompanying Fig. (3.8) shows a simple distribution which can be approximated by a parabola:

$$F(d_i) = a + b(d_i/\bar{d}) + c(d_i/\bar{d})^2.$$

Show that

$$F = \frac{3Nd_i}{2\bar{d}^2}\left(1 - \frac{d_i}{2\bar{d}}\right)$$

where N is the total number of particles. (*Hint*: First show that $F = bu(1 - u/2)$ where $u = d_i/\bar{d}$.) Verify that $\bar{d} = \int d_i \mathrm{d}n_i / \int \mathrm{d}n_i$. What is the maximum value of F?

3.1.6 Calculate the number–volume mean diameter of the particles in Table 3.1 and compare it with the number–area mean diameter calculated in Exercise 3.1.3. Why is it larger? Is this always true?

3.1.7 (a) Show that for the distribution described in Exercise 3.1.5, the standard deviation is given by $\sigma = \bar{d}/\sqrt{5}$.
(b) What fraction of the material lies between $\bar{d} \pm \sigma$ in this case? Compare this with the normal distribution.

3.2 Microscopy

3.2.1 Optical (light) microscopy

The limit of resolution of the optical microscope is a little below 1 μm, and although that can be lowered to perhaps 0.2 μm in the most favourable circumstances, it is not really possible to properly examine colloidal suspensions with the optical microscope. If one is concerned only with the observation of the presence and the motion of particles it is, however, possible to do this using visible light in the **ultramicroscope**. This device makes use of the Tyndall effect, which was mentioned in §2.2.

The method is illustrated in Fig. 3.9. An intense beam of light from an arc lamp or a laser is focused on the particles and they are viewed at right angles to the beam against a dark background, where they appear as tiny pinpoints of light. The amount of light scattered by the particles (§2.2) depends on their volume but also on the relative refractive index, the wavelength of the light and the angle of observation. The minimum size of metal sols which can be seen in the ultramicroscope is about 5–10 nm but for sols of lower refractive index it is rather larger (around 50 nm).

It is also possible to infer something of the particle shape from the ultramicroscope. Particles that are highly anisometric are constantly changing their orientation with respect to the light beam and the observer, as a result of their Brownian motion. This rapid fluctuation in orientation produces a twinkling effect as the scattered intensity varies (rather like stars in the night sky). By contrast, spherical particles show a steady light, although their translational Brownian motion is still readily visible, especially if they are small.

3.2.2 *Electron microscopy*

The most significant development in microscopy has, of course, been the transmission electron microscope (TEM) and, more recently, the scanning electron microscope (SEM). The early colloid scientists were forced to rely on indirect evidence to obtain an idea of particle shape and, in the absence of indications to the contrary, tended to assume that all particles were roughly spherical. There were some exceptions. The crystal structure of the common clay minerals suggested that they should be plate shaped, whilst vanadium pentoxide sol (an important catalyst for sulphuric acid manufacture) showed striking optical and viscous properties which indicated that it was composed of long rods.

Thanks to the electron microscope we can now determine the shape of colloidal particles with very little uncertainty. Solid crystals present little problem since they are largely unaffected by the specimen preparation and by exposure to the electron beam. Some of the softer polymers (like polymethylmethacrylate) do, however, have a tendency to melt in the intense electron beam. Even more difficult to deal with are those submicroscopic structures which are sensitive to the presence of the solvent and likely to be destroyed by the drying process which is essential in electron microscopy before the sample is exposed to vacuum in the beam. The modern techniques developed for preparation of biological specimens have been adapted to colloid systems with great effect.

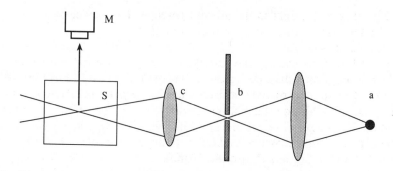

Fig. 3.9. The slit ultramicroscope. An intense beam of light from an arc source or a laser, a, emerges through a slit, b, and is focused by a lens, c, into a chamber containing the colloidal sol. The light scattered from the particles can be viewed through the microscope, M, against a dark background. (The same effect can be achieved using a microscope fitted with **dark-field illumination** using a paraboloid or a cardioid condenser attached to the light source.)

(a) Transmission electron microscope This device depends for its operation on the wave nature of the electron and the fact that electric and magnetic fields of suitable geometry are able to function like lenses to refract, deflect, and focus an electron beam. The ultimate limit of resolution of an electron microscope is determined by the wavelength of the electron but, in practice, one can-seldom resolve much below 1 nm except with very specialized techniques; the limitation in most machines is in the performance of the magnetic lenses and the maintenance of stable magnetic fields (see Exercise 3.2.2). The following description is taken from Silverman *et al.* (1971).

The electron beam is produced by thermionic emission from a tungsten cathode, C, and is accelerated towards an aperture in the anode, A (Fig. 3.10). It is then focused by a lens, L_1, and passes through the sample, which is mounted on a transparent grid, G. Electrons are absorbed or scattered by the specimen in proportion to its local electron density, and the remainder are transmitted. An electromagnetic objective lens, L_2, collects the transmitted electrons and magnifies the image of the specimen 10 to 200 times, onto the object plane of a magnetic projector lens system, L_3, which induces a further magnification of 50 to 400 times as it projects the electrons onto a fluorescent screen, S. There the image may be viewed directly or photographed with a fine-grained film to be enlarged a further 5 to 10 times.

The overall magnification factor ranges from about 100 to 500 000 times. Since the human eye can discriminate between points separated by about 0.2 mm, the lower limit of observation is then from 2 μm down to about 0.4 nm, though specialized procedures can actually resolve individual atoms (0.2 nm) in favourable cases.

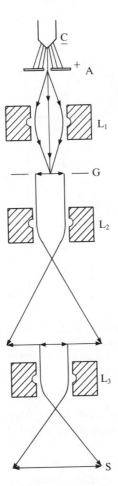

Fig. 3.10. Schematic diagram of the transmission electron microscope. (See text.)

The image formed by this procedure is a two-dimensional representation of the actual structure, and in some cases this is all that is required. In many cases, however, it is helpful to have an idea of the surface topography and this is most readily achieved by **shadow casting**, as illustrated in Fig. 3.11. A beam of metal atoms is fired, *in vacuo*, at an angle to the sample and the deposit modifies its electron transmission characteristics. Figure 3.11 shows schematically how the intensity might be expected to vary in the case shown. Although the human eye will quickly interpret the resulting image in terms of a particle shape it is important to realize that the pattern of light and darkness on the screen or photograph is a result of a complex sequence of events and an exact shape analysis may call for more detailed consideration

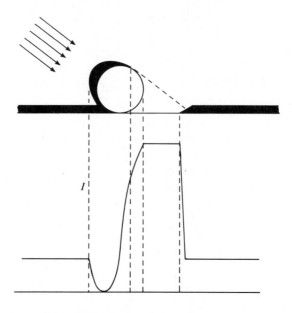

Fig. 3.11. Shadowing (or shadow casting) of a spherical or cylindrical particle produces a characteristic intensity pattern in the transmitted beam, since the shadowing material (usually a metal) is a strong absorber–scatterer of electrons. I is the intensity (or number) of transmitted electrons.

of the influence of shadowing angle θ and direction on the apparent shape. Metals like gold, platinum, palladium, nickel, and chromium may be used for shadowing.

Problems of electrostatic charging, melting, evaporation, and decomposition in the beam can be minimized by careful specimen preparation and the use of techniques first introduced to study biological samples. Of these, the process of replication is the most important in colloid studies. It is illustrated in Fig. 3.12. By using a suitable material to form the replica, a very labile surface can be converted into one of the same shape which will stand up to the electron bombardment. The technique has been used to study surface films and is used in the preparation of the final image in freeze-fracture methods.

(b) Scanning electron microscope Even with the use of replication and shadowing techniques, the transmission electron microscope is limited in its ability to show particle shape; it cannot, for instance, give information about re-entrant surfaces, although taking stereoscopic views can produce something of a three-dimensional effect. The **scanning electron microscope**

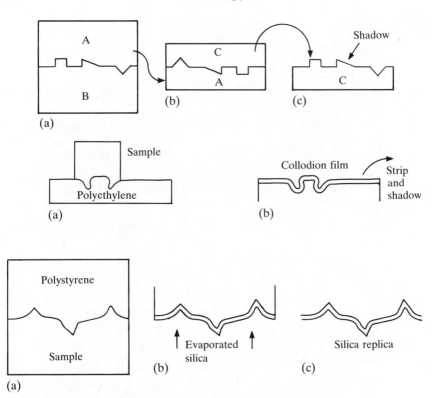

Fig. 3.12. Illustrating various methods of obtaining a replica of a surface. Top: (a) negative replication; (b) positive replication; (c) shadow casting of stripped positive replica. Middle: polyethylene method of positive replication. (a) sample caste in polyethylene; (b) collodion positive replica. Bottom: silica replica method. (a) Negative replication with polystyrene; (b) silica evaporated onto replica surface in vacuum; (c) polystyrene dissolved to reveal positive silica replica. (After Hall (1966).)

(SEM) is able to provide quite remarkable images, which are interpreted by the eye as truly three-dimensional (Fig. 3.13).

In the SEM an electron beam is focused to about 5–10 nm and deflected in a regular manner across the surface of the sample, which is held at an angle to the beam. The low-velocity secondary electrons that are emitted as a result are drawn towards a collector grid and fall onto a sensitive detector. The output from this detector is used to modulate the intensity† of an electron beam in a cathode ray tube (CRT). The beam itself is made to scan the surface of

† To 'modulate the intensity' means to change it at each point in the sweep in proportion to the size of the signal coming from the detector.

Fig. 3.13. Scanning electron micrograph of small (about 0.1 μm) polystyrene (PS) particles adsorbed onto larger PS particles (about 1 μm). (Photograph courtesy of Professor Brian Vincent, University of Bristol.)

the CRT in synchronism with the scanning of the sample by the primary electron beam. The result is a reconstructed image on the CRT much like a television picture.

The big advantage of the SEM is that the secondary electrons are emitted at a low voltage and so can be easily deflected to follow curved paths to the collector. The electrons emerging from parts of the surface that are out of the line of sight are also collected (though at lower intensity) and it is this that contributes most to the striking realism of the three-dimensional image. The *depth of field* is also very large (300–500 times bigger than for a light microscope) so that many different levels of the object can remain in focus at the same time. For this reason the SEM is often used to examine the fine detail of quite large objects (like the body features of insects).

(c) Atomic force microscope   This very recent development in effect draws a relief map of the surface by holding a probe at a fixed distance from the surface as the probe tip moves across it. The distance through which the probe must be raised or lowered at each point on the surface traces a picture of the surface shape in great detail. The probe is maintained extremely close to the surface (about a nanometre or less) and the success of the device depends on its ability to maintain that spacing with great precision and to measure accurately the motion of the tip with respect to a fixed reference height.

Exercises

3.2.1 Show that the average radius of spherical particles is given by:

$$\bar{r} = \left(3CV/4\pi\rho N\right)^{1/3}$$

where ρ is the particle density, C is the concentration of the sol (by mass) and it is found to contain N particles in a volume V. This relation can, in principle, be used to estimate particle size from measurements of N/V using an ultramicroscope.

3.2.2 The wavelength associated with an electron is given by the de Broglie relation, $\lambda = h/p = h/(2m_e E)^{1/2}$, where h is Planck's constant, p is the electron momentum, m_e its mass, and E its kinetic energy. Estimate the wavelength of an electron that has been accelerated through a voltage of 10 kV so that it has acquired an energy of 10 keV. (A typical microscope acceleration voltage would be of this order)

3.3   Sedimentation methods

The determination of particle size by direct microscopic observation has some drawbacks. To begin with the sample must be dried out but, more importantly, only a tiny fraction of the material can be examined. There are also many situations in which one would like to be able to examine the size continuously, for example in a production line. For these reasons there are a large number of indirect methods for determining the particle size, some of which have been in use for a long time and some of which represent the latest technology.

Of these indirect methods the oldest, and one of the most widely used, depends on the rate of sedimentation of particles in a gravitational or centrifugal field.

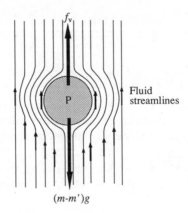

Fig. 3.14. Forces on a particle settling under gravity.

3.3.1 Sedimentation under gravity

When a particle of mass m begins to settle through a fluid under the influence of gravity (Fig. 3.14) it is initially acted upon by three forces: the gravitational force, mg, the upthrust due to the displaced fluid, $m'g$, and the frictional force, f_v, due to the viscous drag of the surrounding fluid. From Newton's law the net force is given by

$$mg - m'g - f_v = m \, du/dt \qquad (3.11)$$

where g is the acccleration due to gravity and u is the velocity at time t.

The frictional force increases with the particle velocity and for colloidal particles settling in a dense medium (like water) it very quickly balances the net downward force (Exercise 3.3.1); the acceleration is then zero and the particles travel with their **terminal velocity**, u_t. For a particle of reasonably regular shape, the drag force is, as we noted earlier

$$f_v = Bu_t. \qquad (2.9)$$

For a rigid spherical particle of radius r we can again set

$$B = 6\pi\eta r. \qquad (2.13)$$

Thus if ρ_s and ρ_l are the densities of the solid particles and the liquid respectively, we have, for the **Stokes settling radius**, r

$$(\rho_s - \rho_l)\frac{4}{3}\pi r^3 g = 6\pi\eta u_t r \qquad (3.12)$$

so that

$$u_t = \frac{2(\rho_s - \rho_l)gr^2}{9\eta}.$$ (3.13)

Although, in principle, u_t can be measured directly with an ultramicroscope it is more usual to follow the changes in particle concentration at a certain depth, h, below the surface of the suspension. After time t all of the particles for which $u_t(r) \geq h/t$ will have settled beyond this depth and, by following the concentration as a function of time, a particle size distribution can be built up.

For nonspherical particles it is not strictly valid to use eqn (2.13) for the frictional coefficient but we have already shown that when Brownian diffusion is large enough to be measured, the frictional coefficient can be calculated directly from Einstein's equation:

$$B = kT/D.$$ (2.12)

When that occurs, however, the settling rate may be significantly affected by the thermal diffusive motion of the particles. The r.m.s. displacement for a sphere is (from eqns (2.12), (2.13) and (2.17)):

$$(\bar{x}^2)^{1/2} = (kT/3\pi\eta r)^{1/2}\sqrt{t}$$ (3.14)

which, for a 1 μm particle in water would be about 0.7 μm in 1 s. The gravitational settling by a particle of density 2×10^3 kg m^{-3} in the same time would be only about 2 μm.

For colloidal particles, then, gravitational settling is of limited use, except for very dense particles. Apart from the interference of Brownian motion, there are considerable practical difficulties in ensuring the necessary degree of temperature stability over the long settling periods; very small convection currents can easily upset the results. Settling is, however, widely used in industry and agriculture for characterizing suspensions of fairly dense solids towards the upper end of the colloid size range. Even if the particles are not spherical and they are too large to have a measurable Brownian diffusion coefficient, one can still obtain an estimate of the 'equivalent settling radius', r_e, which, though difficult to interpret exactly, is undoubtedly a useful comparison measurement between different samples of the same material. For any particular particle, r_e is the radius of the sphere of the same density which settles at the same rate.

3.3.2 Centrifugal sedimentation

The time required, even for a large colloidal particle, to settle through a reasonable distance under the influence of gravity alone (Exercise 3.3.2) makes that procedure rather limited. In most cases it is necessary to increase the sedimentation rate by subjecting the particles to centrifugation. Apart

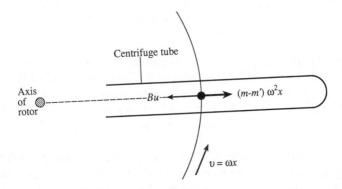

Fig. 3.15. Forces on a particle in a centrifuge tube. The outward force is an apparent (virtual) force invoked to explain the motion of the particle with respect to a coordinate frame attached to the rotor and moving with it.

from the saving in time, there is then less danger of convection currents upsetting the results and the distance moved by sedimention can be much greater than the Brownian motion.

Consider a particle immersed in a liquid at a distance x from the axis of the centrifuge head (or **rotor**) (Fig. 3.15). If it were to stay at that distance as the head revolved, it would have to be acted on by a **(centripetal)** force directed towards the centre of the rotor and forcing the particles to travel in a circle. The magnitude of that force is equal to $(m - m')\omega^2 x$, where ω is the angular velocity (rad s^{-1}) of the rotor and $m - m'$ is the apparent mass (corrected for bouyancy) of the particle. In the absence of that force, the particle moves away from the axis of rotation as if it were acted on by a force of that magnitude acting outwards. Again it is retarded by a frictional force that is proportional to its velocity and again it takes only a short time before these two forces are balanced (Exercise 3.3.1):

$$(m - m')\omega^2 x = Bu(x) = B\frac{dx}{dt}. \tag{3.15}$$

Note, however, that in this case the velocity is not constant but increases as the particle moves towards the outer end of the tube. The quantity:

$$S = \frac{u(x)}{\omega^2 x} = \frac{m - m'}{B} \tag{3.16}$$

is called the **sedimentation coefficient** and is an important characteristic of the material.

For polymeric materials (including proteins) a more appropriate form is obtained by considering the molar mass, M, of the solute so that

$$m - m' = \frac{1}{N_A}(M - \bar{V}_2\rho_1) = M(1 - \bar{v}_2\rho_1)/N_A$$

where $\bar{V}_2$ is the (partial) molar volume of the polymer and $\bar{v}_2$ is the volume per unit mass $(= \rho_s^{-1})$; N_A is the Avogadro constant. The sedimentation coefficient can then be written (using eqn (2.12)):

$$S = M(1 - \bar{v}_2\rho_1)/N_A B = \frac{M\mathfrak{D}}{RT}(1 - \bar{v}_2\rho_1) \qquad (3.17)$$

and, hence, a knowledge of S and the diffusion coefficient, $\mathfrak{D}$, can be used to estimate the molar mass (or molecular weight) of the polymer.

S is obtained by writing eqn (3.15) in the form

$$\frac{dx}{x} = \frac{m}{B}\left(1 - \frac{\rho_1}{\rho_s}\right)\omega^2 dt = S\omega^2 dt$$

and on integration

$$\ln\frac{x_2}{x_1} = S\omega^2(t_2 - t_1). \qquad (3.18)$$

Plots of $\ln x$ as a function of time at known rotation speed can therefore be used to determine S. For colloidal particles where the notion of molar mass is inappropriate, the apparent mass of the particle can be obtained using eqn (3.16) and the friction factor estimated from diffusion experiments (eqn (2.12). Of course if the particles can be assumed to be spherical one may use eqn (2.13) to estimate B directly and hence determine the radius:

$$r = \left(\frac{9\eta S}{2(\rho_s - \rho_1)}\right)^{1/2}. \qquad (3.19)$$

Values of S range from about $1\ \mu s$ for large colloidal particles down to values of the order of $10^{-13}\ s$ for proteins.

The simplest application of eqn (3.18) is in the **two-layer sedimentation** technique in which the colloidal suspension is initially placed in a thin layer on top of the clear suspension medium. As centrifugation proceeds, the components of that layer travel down the centrifuge tube at rates determined by eqn (3.18) and, in favourable cases, their individual progress can be followed, usually by measuring the light absorption at different depths.

3.3.3 Instrumentation

Although most modern instruments use some centrifugation to speed the settling rate, the Sedigraph addresses the problem by a different method. In that instrument, the concentration of particles at a particular level is measured by determining the optical density. This measurement is then repeated at successive higher levels, so that the detection device is moving upwards against the direction of sedimentation.

Calculation of the size distribution is more complicated in the case of centrifugation, because the particle velocity varies with distance from the rotor axis as well as with size. One of the ways around this is to start the material at a point far from the axis of rotation and allow it to sediment through a short distance. This is the procedure used in the **disc centrifuge.**

More modern instruments use more elaborate computer programs to allow a more complex pattern of sedimentation to be followed. In the Horiba Capa 700, for example, the instrument can be programmed to allow a short gravity settling period, to remove large particles, followed by a period in which the rotor speed is gradually accelerated at a certain rate, so that the rotor speed is never held constant. This enables a very wide range of particle sizes to be treated in a single run. The complexities of the settling pattern can be handled by the computer software and are not obvious to the operator.

Exercises

3.3.1 Consider a spherical particle settling under gravity according to eqn (3.11). Show that during the period before it reaches terminal velocity its equation of motion is

$$du/dt = \rho g - Gu$$

where $\rho = (\rho_s - \rho_l)/\rho_s$ and $G = 9\eta/2\rho_s r^2$. Hence show that $u = (\rho g/G)$ $[1 - \exp(-Gt)]$ during this period. How long does it take for a particle of radius 1 μm and density 3 $\times$ 10^3 kg m^{-3} to reach 99% of its terminal velocity in water (density = 10^3 kg m^{-3}, viscosity = 10^{-3} N m^{-2} s)? Repeat the calculation for a radius of 0.1 μm and 0.01 μm.

3.3.2 A suspension of silica particles ($\rho = 2.8$ g cm^{-3}) in water is allowed to settle in a cylinder at 20 °C. Calculate the time required for a particle of radius 2 μm to settle a distance of 20 cm, assuming it is spherical. (Take $\eta = 1$ centipoise = 10^{-2} g cm^{-1} s^{-1}.) Convert the data to SI units and repeat the calculation.

3.3.3 Derive eqn (3.19).

3.3.4 The time taken for a particle to reach its terminal velocity is about $5/G$ (Exercise 3.3.1). Show that the sedimentation coefficient, S, is also an approximate measure of this time.

3.4 Electrical pulse counting

One of the best known electrical pulse counters is the Coulter counter, which is able to count the number of particles in a known volume of solution (about 5% salt) by drawing the suspension through a very small hole that has an electrode on either side of it. Passage of a particle through the hole interferes

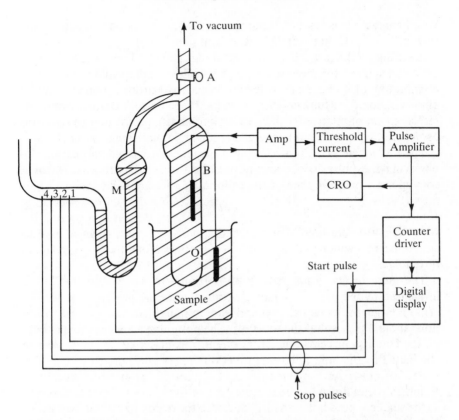

Fig. 3.16. Schematic arrangement of the Coulter counter. (After Allen (1975).)

with the current flow between the electrodes and so the resistance changes. The resulting change in current can be amplified as a voltage pulse and the number of pulses generated by the sample can be counted. A schematic diagram of the apparatus is shown in Fig. 3.16. With a hole of accurately known diameter, a stable current supply and a known (variable) amplification factor, the height of each current pulse becomes proportional to the volume of the particles travelling through the hole, provided that they go through one at a time. Using an electronic **pulse height analyser** one can then convert the device into a counter and sizer.

A mercury manometer, M, on the left is attached to a gentle suction apparatus, which creates sufficient vacuum to draw some suspension through the hole, O, drilled in a flat sheet of sapphire set into the tube, B. It also unbalances the mercury column, drawing it back to the right of the first electrical contact (1). The instrument display sets itself to zero and the operator then closes the connection to the vacuum (tap A). The mercury syphon then takes over and draws more suspension in through the hole.

When contact (1) is reached, counting starts and it can be stopped after a flow of 0.05 ml (contact 2), 0.5 ml or 2 ml (contact 4).

A setting on the instrument (the **threshold**) allows all pulses lower than a certain height to be ignored so that by advancing the threshold a cumulative distribution of pulse heights (proportional to particle volume) can be obtained. Some instruments have an upper threshold, which makes possible measurement of relative frequency distribution (that is the number fraction of material lying between two predetermined volume limits).

The instrument is calibrated using dispersions of known particle size by observing the pulse height and noting the threshold setting to which it corresponds. If the volume of the calibrating particles is v_c, then:

$$v_c = kt_c \qquad (3.20)$$

where t_c is the threshold. The calibration constant k can then be used to determine the volume of an unknown particle from the corresponding threshold.

The pulse height is not strictly proportional to particle volume but is influenced by the size of the particle compared to the size of the hole. The effect is larger for larger particles and leads to some distortion in the distribution (about 5 per cent at the top end). The pulse shape also has some effect on the counting process but errors due to this effect can be detected and eliminated by the built-in editing function. Tubes are supplied with a wide variety of hole sizes ranging from a minimum of 16μm up to more than 1 mm. Particles down to about 5 per cent of the hole size can be measured so the smallest-sized hole can just reach into the region of colloid sizes below 1μm. The instrument is, however, much better suited to measurements above the colloid size range (from a few micrometres up).

3.5 Light scattering methods

We noted in §2.2 that light scattering offered a number of opportunities for determining the size of colloidal particles. The method chosen depends on the size of the particles in relation to the wavelength of light. The simple sort of (Rayleigh) scattering which is described in §2.2 applies only for very small particles, of a few nanometres radius. In this region the scattered intensity depends on the particle volume (eqn 2.21)) and for such very small particles (proteins or synthetic polymers) the volume can be related to the molecular weight or molar mass. As the size increases the notion of molar mass becomes less relevant and, at the same time, the scattering pattern becomes progressively less symmetric, with more emphasis on forward scattering ($\theta < 90°$) compared to back scattering as the size increases. This effect can be used to estimate size by determining the **dissymmetry ratio** which is the ratio of light scattered at 45° to that scattered at 135°. At still larger sizes (around 1 μm) the scattering pattern becomes extremely complicated even for a sphere but the theory has been calculated (Fig. 2.6) and in principle it can be used to determine the size exactly. At still larger sizes, above about 10 μm, the situation becomes simplified again as the particles are now much larger that the wavelength and they **diffract** the light rather than scattering it. Small particles diffract the light through large angles, whilst large particles cause less diffraction and again this can be used to determine the size of particles up to about 1 mm quickly and efficiently.

It is therefore the upper end of the colloid range which poses the greatest computational problem, but with the advent of fast computer processing, this can be turned to advantage. We also have the possibility of using the alternate **dynamic light scattering** in this region because it is here (for $r < 1\,\mu$m) that the particles show significant Brownian motion and a measurable diffusion coefficient.

3.5.1 Intensity methods

(a) Rayleigh scattering We noted in §2.2 that, even in the Rayleigh scattering region ($d \ll \lambda$) the amount of light scattered at any angle was a strong function of particle size but was symmetrical about the $\theta = \pi/2$ direction. The calculation of the size, or the molar mass (molecular weight) of the scattering entity uses eqn (2.19) with the polarizability, α, estimated from its refractive index, n, as suggested by eqn (2.20), but in the form

$$\alpha = \varepsilon_0 \frac{n_1^2 - n_0^2}{N} \tag{3.21}$$

where n_0 is the refractive index of the surrounding medium and N is the number of scatterers per unit volume.

We now define a quantity called the **Rayleigh ratio**, R_θ:

$$R_\theta = \frac{Ir^2}{I_{0,u}} = \frac{\pi^2}{2\lambda^4 N}(n_1^2 - n_0^2)^2(1 + \cos^2\theta)$$

$$= R_{90}(1 + \cos^2\theta) \tag{3.22}$$

where R_{90} is the value of R_θ evaluated at right angles to the incident beam.

We can write

$$n_1^2 - n_0^2 = (n_1 - n_0)(n_1 + n_0) \approx 2n_0(n_1 - n_0) \tag{3.23}$$

and

$$n_1 - n_0 = h_1 C \tag{3.24}$$

where h_1 is a constant and $C = NM_i/N_A$ is the concentration of molecules (mass per unit volume). Equation (3.24) expresses the assumption, which can readily be verified by experiment, that for dilute solutions the refractive index increase is proportional to the amount of solute. Substituting eqns (3.23) and (3.24) in (3.22) gives

$$R_{90} = k_1 C M_i \qquad (3.25)$$

where

$$k_1 = 2\pi^2 n_0^2 h_1^2 / \lambda^4 N_A. \qquad (3.26)$$

Measuring the amount of light scattered at an angle of 90° to the incident beam allows one, in principle, to estimate the molar mass of a polymer in very dilute solution. The instrument used is called a **nephelometer**. An alternative procedure, used for more concentrated systems, is to measure the **turbidity**, T, of the sample. This is the measure of the reduction in intensity of light as it passes through the sample, due to the scattering process. The fraction of the incident light scattered in all directions by a collection of particles is obtained by integrating the angular intensity function $I(\theta)$ over the surface of a sphere and dividing by the initial intensity. That light is removed from the transmitted beam as though it were absorbed by the solution. The turbidiy (Exercise 3.5.1) is given by

$$T = (16\pi/3)R_{90}. \qquad (3.27)$$

Note that this has dimensions of reciprocal length (since R_{90} is proportional to $(\lambda^4 N)^{-1}$) because R_{90} measures the amount of scattering occurring from a unit path length of the suspension.

As the beam passes through the suspension, this fraction of light is being lost at every point in the path. The reduction in the light intensity is given by

$$-dI/dx = TI$$

and the fraction of light transmitted through the path length, l, is obtained by integration (Exercise 3.5.1):

$$I_t/I_0 = \exp(-Tl). \qquad (3.28)$$

The turbidity is, thus, analogous to the extinction coefficient as measured by the Beer–Lambert law (Atkins 1982, p. 605). Measuring either T or R_{90} allows one to estimate the molar mass from

$$M_i = \frac{R_{90}}{k_1 C} = \frac{T}{HC} \qquad (3.29)$$

where

$$H = \frac{32\pi^3}{3} \frac{n_0^2}{\lambda^4 N_A} h_1^2. \qquad (3.30)$$

A more elaborate treatment, given by Debye, shows that for reasonably dilute solutions a more accurate formulation is

$$T = \frac{HC}{(1/M_i) + 2BC} \qquad (3.31)$$

where B is the **second virial coefficient** of the scatterer (Atkins 1982, p. 792). Equation (3.29) is obviously the limiting form of (3.31) as C or B tends to zero. A plot of HC/T against C thus has an intercept of $1/M_i$ and a slope equal to twice the virial coefficient (which is itself a useful measure of the interaction between the molecules).

In order to treat Rayleigh scatterers as particles it is best to return to the fundamental expression eqn (2.21) in terms of the particle volume, v. The turbidity can then be expressed relative to the volume fraction $\phi \, (= N_p v)$ as (Exercise 3.5.2):

$$T/\phi = 24\pi^3 \left(\frac{n^2 - 1}{n^2 + 2} \right)^2 \frac{v}{\lambda^4} = H' v \qquad (3.32)$$

where n is the relative refractive index $(= n_1/n_0)$. Notice that in eqn (3.32), the turbidity increases with the size of the particles even when the volume fraction, (or the mass fraction) remains constant. This provides a very convenient method for following the kinetics of coagulation of a colloidal sol or for checking the stability (in the colloid sense (§2.3)) of a suspension.

(b) Rayleigh–Gans–Debye (RGD) scattering When the particles become somewhat larger (10 to 100 nm say), so that the scattering pattern loses its symmetry, it becomes possible to estimate the particle size by examining some features of the intensity pattern as a function of the angle of scattering. The theory is more complicated but suitable approximations were worked out by Gans and Debye. Some discussion of the theory is given by Atkins (1982, p. 831) so we need only note that the expression for R_θ is modified to take more account of the angle:

$$R_\theta = P(\theta) \, (R_\theta)_{\text{Rayleigh}} \qquad (3.33)$$

where the scattering factor $P(\theta)$ is a consequence of the interference which occurs between light waves scattered from different parts of the same particle. The calculated values of $P(\theta)$ for different shapes are shown in Fig. 3.17. If the shape is known, a better estimate of the molar mass can be made in the RGD regime by reference to Fig. 3.17.

In the RGD region it is also possible to estimate the radius of monodisperse spherical particles simply by measuring the so-called dissymmetry ratio:

$$R_D = I_{45°}/I_{135°}.$$

Determination of particle size

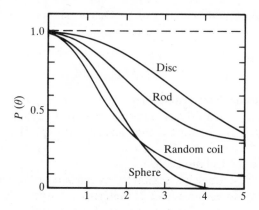

Fig. 3.17. Scattering function $P(\theta)$ for macromolecules of various shapes. (After Marshall (1978).) The abscissa represents $Qd/2n_0$ where Q is the scattering vector and d is the diameter for the sphere or disc, and the length for the rod. For the random coil, the abscissa represents $Qd/\sqrt{6}n_0$ where d is the r.m.s. end-to-end distance.

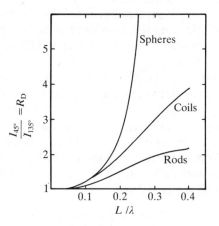

Fig. 3.18. The dissymmetry ratio R_D as a function of (relative) characteristic length for spheres, random coils, and rods. (After Kerker (1969).)

As we noted earlier, as the size increases, the scattering tends to increase in the forward direction at the expense of back scattering, as can be seen from Fig. 3.18.

Alternatively one can use the theoretical form of $P(\theta)$ to assess the particle size. For particles of arbitrary shape obeying the RGD criterion, the limiting form of $P(\theta)$ for small scattering angles is

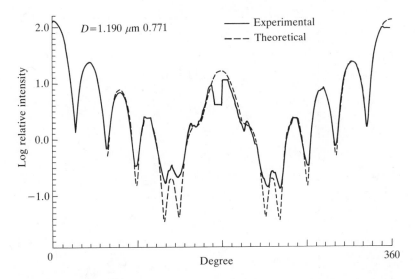

Fig. 3.19. Comparison of experimental data with (Mie) theory for scattering from a single polystyrene sphere. 0.771 is the least squares deviation between theory and experiment. (After Marshall *et al.* (1976).)

$$P(\theta) = 1 - \frac{(Qa_G)^2}{3} + \dots \tag{3.34}$$

where Q is the scattering vector (eqn (2.23) and a_G is the radius of gyration of the particle. Equation (3.34) can be used for particles with radius in the range 20–100 nm. It is not difficult to show that a plot of $\ln I_\theta$ as a function of Q^2, called a **Guinier plot,** will have an initial slope $-a_G^2/3$. This procedure has become much more useful with the advent of lasers because they permit measurements at very small scattering angles. The procedure is called LALLS (low-angle laser light scattering).

(c) Mie scattering When the particles are comparable in size or a little larger than the light wavelength, the scattering pattern becomes extremely complicated. The theory had been worked out in the early 1900s and as Fig. 3.19 shows it is remarkably accurate in describing the details of the pattern. Until recently the calculation was a major problem but with the advent of fast computer processing it is now possible to turn that complexity to advantage. For a system in which the particles can be assumed to be close to spherical it becomes possible to determine the size distribution by matching the observed scattering pattern with that expected from particles of different sizes. Figure 3.20 shows how successful the procedure can be.

To get data of the necessary precision, the scattering angle must be

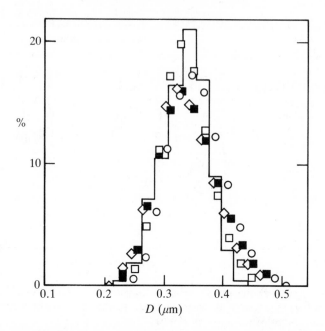

Fig. 3.20. Comparison of size distribution of vanadium pentoxide aerosol with electron microscope result (shown as histogram). Wavelength (λ in nm): ■ 406; ○ 436; □ 546; ◇578. (From Kerker (1969, p. 367).)

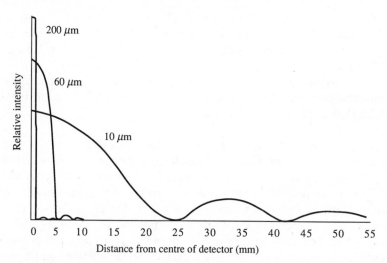

Fig. 3.21. Distribution of light intensity due to diffraction of light by relatively large particles. Note that small particles scatter to much larger angles than large particles do. (After McFadyen, Malvern Instruments.)

accurately known and that is best achieved with a **goniometer** arrangement (Atkins 1982, p. 740) to collect light at a precise angle to the incident beam.

For the more varied shapes that occur in real colloidal suspensions it pays to be a little sceptical about the distributions which the computer so readily generates. In most cases the quoted mean value will be a useful measure of some aspect of the suspension size and the program will also give a reasonable estimate of the spread of data about the mean. The details of the distribution may, however, be quite misleading; small errors in the data can upset the algorithms used to extract the distribution from the scattering data.

(d) Fraunhofer diffraction For somewhat larger particles (above about 10 μm) the scattering pattern becomes much simpler (Fig. 3.21), the scattering angle is smaller and the angular pattern does not need to be determined with the same precision as that for the Mie scatterers. In this regime it is more important to be able to collect data from many different angles as quickly as possible. The solution used in the Malvern instrument is to illuminate the particles with a laser beam and to collect the diffracted light with a special lens which directs it to an array of detectors (Fig. 3.22). The computer can then determine the angular distribution of light from the output of the array. This technique is best suited to particles above the colloid size range.

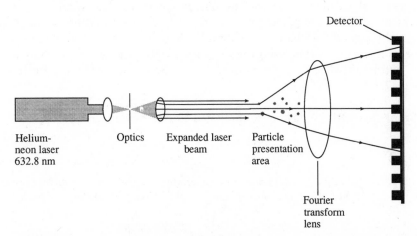

Fig. 3.22. Schematic arrangement of the optical system in the Malvern 3600Ec light scattering apparatus. In the MasterSizer the system is reversed: the sample is confined in a narrow tube and the Fourier lens is interposed between the laser and the sample. In that case the detector array is moved in all three directions to obtain a good sample of the scatter pattern.

For particles around 1 μm or less, the scattering angles are much larger and a more elaborate method is required to collect the data. In their MasterSizer, Malvern use a moveable array to cover the necessary scattering range. An alternative procedure is to use the **dynamic** light scattering method which we will now discuss.

3.5.2 *Dynamic light scattering (QELS or PCS)*

The basic principle of this method was discussed in §2.2.2. In its original form the method involved using a spectrum analyser to determine the intensity of the light scattered at a certain angle as a function of the frequency as shown in Fig. 2.7. The diffusion coefficient could then be estimated from the broadening of the line due to the Doppler effect.

In more recent applications of the technique an alternative procedure called **photon correlation** is used. The idea is to estimate how much particle motion is occurring by measuring the signal at a certain instant in time and then comparing it with signals obtained at successively later times. By multiplying the initial signal by the one obtained at a number of later times and summing (or integrating) the result one obtains a measure of how highly **correlated** the signal is. In the extreme case where there is perfect correlation the size of the product signal will not change with time, but if the particles are moving rapidly then the signal collected a little while after the initial signal will show no similarity to it. The correlation will be poor and the product signal will approach zero. The measurements are taken at time intervals which are short compared with the time for diffusive motion, so there is initially a good correlation which gradually disappears as the time interval becomes longer. By analysing the exact shape of this decay curve it is possible to extract the diffusion coefficient, D, and hence to obtain the size from eqn (2.18).

A number of commercial instruments (for example the Malvern and the Brookhaven BI-90) have been marketed to take advantage of this effect.

Exercises

3.5.1 The turbidity, T, is determined by integrating the scattered light intensity over the surface of a sphere surrounding a point in the path of the incident light:

$$T = \frac{1}{I_{0,\,u}} \int_{\theta=0}^{\pi} \int_{\phi=0}^{2\pi} I(\theta) r^2 \sin\theta \, d\theta \, d\phi.$$

Verify eqns (3.27) and (3.28).

3.5.2 Show that, for a polymer solution scattering in the RGD region,

$$\frac{KC}{R_\theta} = \left(\frac{1}{M} + 2BC\right)\left(1 + \frac{16\pi^2 a_G^2}{3\lambda^2} \sin^2(\theta/2)\right).$$

λ in this case is the wavelength of the light *in the medium* $(\lambda = \lambda_0/n_0)$. This relation is the basis of the **Zimm plot** which is used to determine the molar mass of a polymer from measurements at moderate concentration where interaction between the polymer molecules cannot be neglected. The method depends on plotting KC/R_θ against $\sin^2(\theta/2) + kC$ where k is an arbitrary constant chosen for convenience. M is obtained by extrapolating to zero C and zero θ. (See Hiemenz (1986) for details.)

3.5.3 Establish eqn (3.32).

3.6 Hydrodynamic methods

3.6.1 Capillary hydrodynamic fractionation

Another recent development in the sizing of (particulate) colloidal dispersions is the use of capillary fractionation in which the suspension is forced, under pressure, either through a long column packed with porous beads or through a long, fine capillary tube. Both procedures depend on the same theoretical ideas. Particles of different size travel at different speeds through the column or tube and the effluent can be collected in fractions, as in chromatography. The particle concentration can be determined in each fraction, by measuring its light absorption characteristics. It turns out that large particles travel through the system faster than small ones because the smaller particles spend more time near the walls of the capillaries (Fig. 3.23) where the flow rate is slower; the centres of larger particles cannot approach those parts of the tube and tend to spend more time near the tube axis.

The initial application of the method forced the liquid dispersion under a pressure of about 20 atm through a long column packed with nonporous glass beads of radius about 10 μm. This method was called **hydrodynamic**

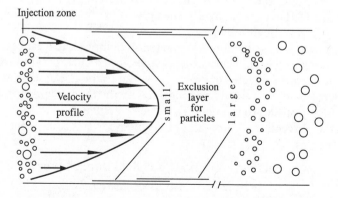

Fig. 3.23. Illustration of the velocity profile and the separation of large particles from small particles in a capillary tube. (After DosRamos *et al.* (1990).)

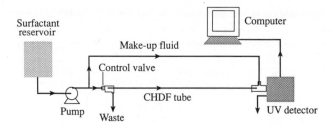

Fig. 3.24. Schematic diagram of the configuration of pump, split flow, and by-pass arrangement in the capillary hydrodynamic flow apparatus (Matec Applied Sciences, Massachusetts, USA). The flow is split so that the amount of material flowing through the detector system can be increased, because flow through the CHDF tube is so restricted.

chromatography by its originators but since it does not involved the partitioning of the colloid between two phases (as occurs in other forms of chromatography) the more recent name for it is **capillary hydrodynamic fractionation** (CHDF).

The column method was easy to implement, since small glass beads are readily available, but it suffers from the fact that colloidal particles become adsorbed on the beads and block the column. It is now possible to obtain capillaries in lengths from 1 to 20 m with very fine bore (4–60 μm internal diameter) and these form the basis for a modern separation and measuring instrument.

The method of operation of the Matec CHDF instrument is shown in Fig. 3.24. The CHDF tube is a few micrometres in diameter and a couple of metres long so its total capacity is only about 30 nl which creates a problem: how to get sufficient fluid through the system to make a measurement. The solution is to split the fluid stream into two parts, only one of which goes through the CHDF capillary whilst the bulk of the fluid (without any added particles) goes through a by-pass.

From a reservoir, a solution of surfactant is pumped into a capillary which splits it into two streams. One stream passes into a chamber into which the unknown sample is injected and then flows out to the CHDF capillary tube. The other stream is returned and remixed with the fluid coming through the CHDF tube just before it enters the detector system.

As in the conventional chromatography methods, it is convenient to define an R_F value as

$$R_F = \frac{\text{Rate of transport of colloid through tube}}{\text{Rate of transport of carrier fluid}}. \tag{3.35}$$

Unlike the normal situation in chromatography the values of R_F in CHDF are normally greater than unity, indicating that the colloidal particles get through the column *faster than the carrier liquid* (or **eluant**). An approximate relation for R_F can be derived as follows. The slow (laminar) flow of a liquid in a cylindrical pore (Fig. 3.23) obeys the Poiseuille equation (compare eqn (4.13)):

$$v(r) = \frac{\Delta P r_0^2}{4\eta l}\left[1 - \left(\frac{r}{r_0}\right)^2\right] \tag{3.36}$$

where $v(r)$ is the velocity of the cylinder of liquid at a distance r from the axis, ΔP is the pressure drop, r_0 is the capillary radius and l is its length. The maximum velocity

$$v_0 = (\Delta P)r_0^2/4\eta l$$

occurs on the axis and $v = 0$ at the wall. The average fluid velocity is (Exercise 3.6.1)

$$\bar{v} = \left[\int_0^{r_0} 2\pi r v(r)\mathrm{d}r\right]\left[\int_0^{r_0} 2\pi r \mathrm{d}r\right]^{-1} = \frac{v_0}{2}. \tag{3.37}$$

A sphere of radius a can approach the capillary wall only to within a distance a. Since these spheres cannot sample the slower fluid velocities close to the wall they will move, on average, faster than the fluid. The average velocity of the sphere can be shown to be (Exercise 3.6.2):

$$\bar{v}_p = v_0[1 - \tfrac{1}{2}(1 - \bar{a})^2] \tag{3.38}$$

where $\bar{a} = a/r_0$. We can then write

$$R_F = \frac{\bar{v}_p}{\bar{v}} = 2 - (1 - \bar{a})^2 = 1 + \bar{a}(2 - \bar{a}). \tag{3.39}$$

The rate of transport of the fluid through the tube can be determined using a solute which is readily measured by its optical absorption; it can be assumed to move at the same speed as the fluid. The measured R_F value could then be used to estimate the size of particles using eqn (3.39). More realistically one can use calibration standards to determine how rapidly a sphere of a given size goes through the capillary and measure the unknown with respect to the standard. The procedure is rapid and is able to accurately separate samples with very similar sizes in a bimodal or even trimodal distribution (Fig. 3.25).

3.6.2 Field flow fractionation (FFF)

There are several versions of this method, depending on the nature of the applied field. Figure 3.26 shows the general layout of the apparatus and control system (for more information see Caldwell (1988)). The suspension to

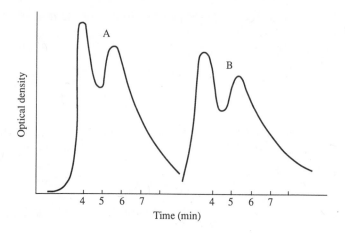

Fig. 3.25. Separation of a bimodal particle distribution using the method of CHDF. Two runs on the same sample illustrating the reproducibility.

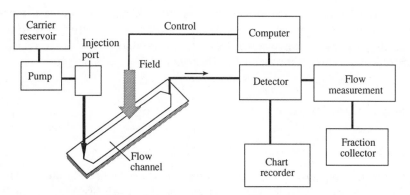

Fig. 3.26. The arrangement used in field flow fractionation methods. The field which is applied at right angles to the flow of the suspension can be gravitational, electric, magnetic, centrifugal, or a cross-flow.

be sized is carried into a thin channel by a carrier fluid and while it is in the channel it is subjected to a steady field of some sort at right angles to the direction of flow. The field could be electrical, magnetic, thermal (that is, a temperature gradient), or a cross flow process.

For example in **sedimentation** FFF the cross-field is a centrifugal force produced in a centrifuge. The principle of the method is similar to that of the other hydrodynamic procedures but the applied field produces a different effect: the larger particles are thrown to the outer regions of the tube (Fig. 3.27) and so tend to move more slowly than the smaller particles which are able to diffuse back towards the axis where the flow velocity is higher.

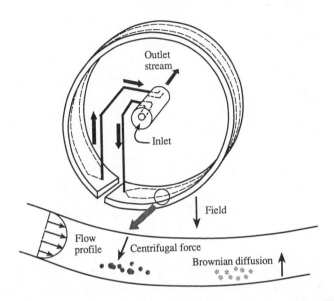

Fig. 3.27. In Sedimentation FFF the flow channel is wrapped around the axis of a centrifuge and the cross-field is provided by the centrifugation process. Small particles move through the system more rapidly in this case because they tend to diffuse back towards the middle of the channel where the flow is faster.

The net result is that smaller particles are eluted *faster* in this case than the larger particles, which is opposite to the behaviour of the CHDF. Very sharp separations have been achieved for mixtures of monodispersed polystyrene latex spheres with sizes in the range 0.2–0.6 μm. The time required for elution may be several hours, but one of the advantages of the method is that the separate sized components come out in fairly pure form so that they can be analysed by other physical or chemical means.

Exercise

3.6.1 Verify eqn (3.37). Why is this average used and not simply $(1/r) \int v \, dr$?
3.6.2 Establish eqn (3.38) by averaging over the inner part of the capillary. Derive eqn (3.39).
3.6.3 The average fluid velocity in the CHDF capillary is about 1 cm s^{-1}. Estimate the pressure drop along its length assuming it is 2 m long and has a diameter of 4 μm.

3.7 Electroacoustics

A very recent addition to the particle sizing methods is provided by the technique of electroacoustics. When a high-frequency sound wave moves

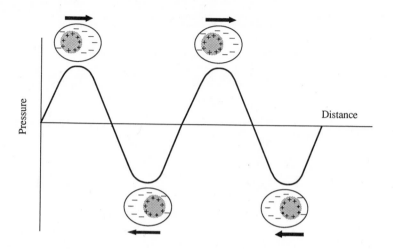

Fig. 3.28. Colloidal particles, with their diffuse double layers, in a sound wave. The wave causes both the particles and their surrounding double layers to move as indicated by the arrows. The larger movement of the charge in the double layer causes a dipole to be set up and the sum of a large number of those dipoles (all pointing in the same direction) generates a measurable potential difference.

through a colloidal suspension it causes the particles to move backwards and forwards at the same frequency as the wave. Even at frequencies of the order of 1 MHz, this particle motion can be detected because it gives rise to an electrical signal. The signal arises because, in addition to the (very small) motion of the particle, there is as larger movement of the ions in the double layer (Fig. 3.28), because the fluid can respond to the pressure wave more quickly than can the particle. (The particle's inertia makes it slower to respond.) This generates a small dipole and the presence of many such dipoles in the suspension, all pointing in the same direction, creates a macroscopic electric field which can be detected by placing two electrodes in the suspension, positioned at the peak and trough in the sound wave. That so-called **colloid vibration potential** has been known for half a century but has been difficult to exploit.

More recently it has been shown that the opposite effect also occurs: if an alternating electric field is applied to a colloidal suspension it generates a sound wave of the same frequency, just like a piezoelectric crystal. That is called the **electrokinetic sonic amplitude** (ESA) effect and it has recently become possible to use it to estimate particle size. The method is based on the fact that there is a phase lag between the applied signal and the resulting response because of the inertia of the particles. The bigger the inertia of the particles, the more difficult it is for them to follow the electrical signal. The ions in the surrounding double layer, however, are able to keep up with

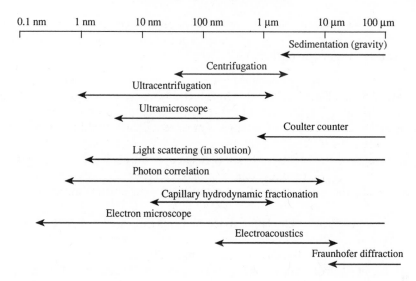

Fig. 3.29. Approximate range of application of the particle size methods discussed in this chapter.

the signal even at 15 MHz. By measuring the phase lag as a function of the frequency of the applied field it is possible to determine the **mass average particle size** of the colloid. The advantage of the method is that it can be applied to quite concentrated suspensions. Optical methods, by contrast, work most effectively when the system is able to transmit light, and that usually means at very low concentrations.

It turns out that the phase lag is often independent of the charge on the particles, so that it is possible to use that phase lag to estimate the particle size and, knowing the size, to estimate the charge on the particles from the *magnitude* of the ESA signal. Since these are two of the most significant features of the colloid, the method is expected to have many applications in the control of colloid chemical systems, especially the slurries that are used in making paper, paints, and inks, and in mineral processing and ceramic manufacture.

3.8 Summary of sizing methods

We have considered particle sizing in some detail for two reasons:

(1) because particle size is an important intrinsic characteristic of a colloid; and

(2) because the practical problem of determining the particle size introduces us to a number of the important theoretical concepts of colloid science.

We have concentrated on methods that can give some idea of the distribution of particle size, rather than just the average, since the latter can be somewhat misleading.

There is another approach which we will examine later, based on the measurement of surface area, which permits an estimate of the surface average radius, $\bar{r}_{NA}$, if one can make some assumption about the particle shape (Exercise 1.1.3). Surface area is most readily measured by an adsorption process. In effect one determines how many molecules of a certain type (the **adsorbate**) are required to completely cover the surface (the **adsorbent**) to a known depth (usually one monolayer). Then if we know the area occupied by the adsorbate molecules, the area of the underlying solid can be estimated.

The range of the methods discussed above is indicated in Fig. 3.29 and it is apparent that there is not a great deal of choice over most of the colloid size range, and even that choice may be limited by other factors. The sampling error in electron microscopy, for example, can be very large if the sample is polydisperse. Particle sizing is a rewarding area for the application of any new procedure if it can be shown to be effective.

References

Allen, T. (1975). *Particle size measurement* (*Power technology series* (ed. J. C. Williams)). Chapman & Hall, London.

Atkins, P. W. (1982). *Physical chemistry*, (2nd edn). Oxford University Press.

Caldwell, K. D. (1988). *Anal. Chem.*, **60**, 959A.

Dos Ramos, J. G. and Silebi, CA. (1990). *Journal of Colloid and Interface Science*, **135**, 165.

Hall, C. E. (1966). *Introduction to electron microscopy*, (2nd edn). McGraw-Hill, New York.

Hiemenz, P. C. (1986). *Principles of colloid and surface chemistry*, pp. 261–5, Marcel Dekker, New York.

Kerker, M. (1969). *The scattering of light and other electro-magnetic radiation*, p. 432. Academic, New York.

Marshall, A. G. (1978). *Biophysical chemistry—principles, techniques and applications*, p. 477. Wiley, New York.

Marshall, T. R., Parmenter, C. S. and Seaver, M. (1976). *Journal of Colloid and Interface Science*, **55**, 624.

Silverman, L., Billings, C. E., and First, M. W. (1971). *Particle size analysis in industrial hygiene*, p. 111 *et seq*. Academic, New York.

4

FLOW BEHAVIOUR

4.1 Deformation and flow

When a material of any sort is subjected to a stress it will be to some extent deformed. The study of the relation between the applied stress and the resulting deformation is called **rheology** and it is an important area of colloid and polymer science. The stress (force per unit area) can be applied in various ways: as a tension, as a compression, or as a shearing process (Fig. 4.1) or some combination of the three.

In compression and tension dilute colloid suspensions behave very much like ordinary liquids. Only in rather special dispersions does one encounter unusual behaviour under tension, whilst under compression most solids and

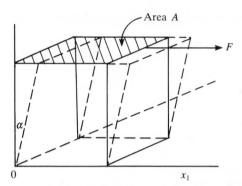

Fig. 4.1. Application of a shearing stress $S(=F/A)$ to a material, produces a strain $\gamma = \tan \alpha$.

liquids tend to behave in a similar way (they just get a little squeezed up). On the other hand, even quite dilute suspensions can show a variety of unusual behaviour patterns under the influence of a shearing stress. Indeed, it is often these unusual deformation patterns which are sought after in the application of a colloidal dispersion. (Getting an ice-cream to have the right consistency or 'feel' on your lips and tongue is a sophisticated exercise in applied rheology.)

Consider the simple shearing regime illustrated in Fig. 4.1. The lower plate is held stationary and the upper plate is pulled with a force, F, acting in the x direction over an area A. The force per unit area applied to the material is called the **shearing stress** and it will cause a deformation or **strain** γ. When the force is removed we find that either:

(1) the material returns to its original shape;
(2) the material remains in its new position, (**flow** has occurred); or
(3) some partial recovery occurs.

These three behaviour patterns are characteristic of solids, liquids and plastics respectively but, in practice, most materials can exhibit any or all of them depending on the time-scale involved in the application of the stress and the measurement of the resulting strain. The shorter the time-scale, the more solid-like the behaviour tends to be. Thus geologists can speak of the 'flow' of rocks over geological time, while a person falling from a great height into water samples its solid-like characteristics when it is deformed over a short time interval.

4.1.1   Ideal solids and liquids

When a tensile (stretching) force is applied to an ideal solid, Hooke's law says that the amount of stretching produced is proportional to the applied stress:

$$F/A \propto \gamma \quad \text{or} \quad F/A = S_\text{T} = Y\gamma$$

where Y is called the **Young's modulus** of the solid. The strain, γ, in that case is the relative change in length and it has no units. The corresponding behaviour under a shearing stress is described in an analogous way:

$$F/A \propto \gamma \quad \text{or} \quad F/A = S = G\gamma \tag{4.1}$$

where G is called the **shear modulus** of the material and γ is defined in Fig. 4.1. Many solids obey eqn (4.1) for small to moderate stresses and, provided the stress remains below some upper limit S_L, they will recover their shape completely when the stress is removed. If $S > S_\text{L}$ the material suffers permanent deformation, that is, some **flow** or **creep** occurs and the solid has begun to exhibit some of the characteristics of a **plastic** or a **liquid**.

Ideal liquid-like behaviour is described as **Newtonian behaviour** and in that case the applied shearing stress is directly proportional to the **time rate of strain** or **rate of shear** ($\mathrm{d}\gamma/\mathrm{d}t$):

$$S \propto \mathrm{d}\gamma/\mathrm{d}t \quad \text{or} \quad S = \eta\,\mathrm{d}\gamma/\mathrm{d}t \quad \text{or} \quad S = \eta\dot{\gamma} \tag{4.2}$$

where the proportionality constant, η, is called the **viscosity** of the liquid and the dot over γ indicates a time derivative. This equation was proposed by Isaac Newton to describe the flow of simple fluids (gases and liquids like water) undergoing steady shear.

In the system shown in Fig. 4.2, a simple liquid is confined between two plates. The lower plate is stationary and the upper plate is being pulled along with a velocity v by the application of a force per unit area S. It can be assumed that the liquid in contact with the lower plate remains stationary and the liquid in contact with the upper plate moves with velocity v. (It is observed experimentally that the liquid in contact with a solid sticks to it so that there is always a layer which moves with the local velocity of the solid.)

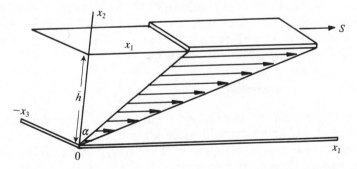

Fig. 4.2. Deformation (flow) of a liquid under an applied shearing stress, S. If the velocity of the upper plate is v then $\dot{\gamma} = \mathrm{d}\gamma/\mathrm{d}t = \mathrm{d}(\tan \alpha)/\mathrm{d}t = h^{-1}\,\mathrm{d}z/\mathrm{d}t = v/h$.

Flow behaviour

Table 4.1 Viscosities of some common liquids (in mPa s or cP) at 15°C

Substance	Viscosity
Water	1.14
Mercury	1.58
Ethyl alcohol	1.34
Carbon tetrachloride	1.04
Olive oil	99
Glycerol	2330

Between the two plates the velocity changes gradually from zero to v. The rate of shear, $d\gamma/dt$ for this simple shear regime is equal to the velocity gradient v/h (or dv/dz) and is normally measured in s^{-1}.

The unit of viscosity in the old c.g.s. system was the poise (1 poise = 1 dyne cm^{-2} s) and in the SI system it is $1\,N\,m^{-2}\,s = 1$ Pascal second (Pa s) = 10 poise. A convenient measure in the old system was the centipoise (1 cP = 0.01 poise) because water has a viscosity of about 1 cP at room temperature. On the new scale the corresponding unit is 1 mPa s. The viscosity of some common liquids is given in Table 4.1, according to which water flows over 2000 times more readily than glycerol.

Although these liquids exhibit a wide range of viscosities, they are all Newtonian in behaviour, that is they obey eqn (4.2) and the viscosity is independent of how fast the liquid is being sheared. Colloidal suspensions, by contrast, often have a variable viscosity, which depends on the rate of shear, and that can be very important. Consider, for example, the way the 'apparent viscosity' of a paint changes during its application. When it is on the brush, the viscosity is high (so it doesn't drip off); it must flow freely when sheared against the surface to be painted, then it must quickly increase in viscosity so that it does not run down (drip or sag) under gravity, but it must flow sufficiently to eliminate the brush marks. The dependence of viscosity on time and the shearing stress to which it is subjected determine the ease with which the paint can be applied.

4.2 Measurement of viscosity

The common methods for measuring viscosity depend on the establishment of a simple steady flow pattern in which it is possibly to estimate both the shear stress and the shear rate. The two most widely used methods are the **Couette** and the **Ostwald** viscometer, and their various derivatives.

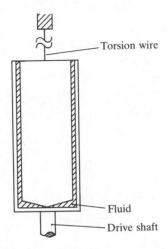

Fig. 4.3. A vertical cross-sectional sketch of a Couette viscometer. The liquid occupies the shaded region between the two cylinders.

4.2.1 Concentric cylinder viscometer

The Couette viscometer consists of two coaxial cylinders, as shown in Fig. 4.3. The space between the two cylinders is filled with the fluid to be measured. One of the cylinders (preferably the outer one) is rotated at a constant rate (Ω rad s^{-1}) and the other is held in place by a torsion wire. The viscous drag of the liquid against the surface of this (inner) cylinder (Fig. 4.4) causes it to twist against the torsion wire and it stops twisting only when the restoring torque in the wire is equal to the torque applied by the moving fluid.†

When the inner cylinder stops turning the flow velocity of the fluid varies from the inner wall ($v(r) = 0$) to the outer wall ($v(r) = \Omega R_o$) where R_o is the radius of the outer cylinder. At first sight one might be tempted to set the shear rate equal to $dv(r)/dr$ but that is not so. We must correct for the fact that some of the rotational motion is a rigid body motion which is not shearing the liquid at all. That is measured by ω, which varies from 0 to Ω across the gap. We then have

$$\text{Shear rate } \dot{\gamma} = dv/dr - \omega = dv/dr - v/r = r(d\omega/dr). \qquad (4.3)$$

The stress on each cylinder of fluid is given by eqn (4.2):

$$S = \eta(dv/dr - v/r) \qquad (4.4)$$

---

† A torque is a twisting force, the size of which is measured by the product of the force and the distance that the force is from the centre about which the twisting motion is being applied.

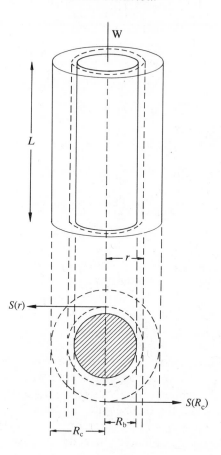

Fig. 4.4. Horizontal section through a Couette viscometer.

and this stress imparts a torque to the cylinder given by:

force × radius = stress × area of surface × radius = $2\pi r^2 LS$

where L is the length of the cylinder. The torque on every cylinder of fluid is the same and is also the same as the torque T_i on the inner cylinder:

$$2\pi r^2 \eta \left(\frac{\mathrm{d}v}{\mathrm{d}r} - \frac{v}{r}\right) L = T_i. \qquad (4.5)$$

This expression can be integrated with respect to r (Exercise 4.2.1) to give

$$v = \frac{T_i r}{4\pi\eta L} \left(\frac{1}{R_i^2} - \frac{1}{r^2}\right) \qquad (4.6)$$

where R_i is the radius of the inner cylinder. On the outer cylinder, v must be equal to $R_o \Omega$ and substituting these values in eqn (4.6) gives for the torque:

$$T_i = \left(\frac{4\pi L R_i^2 R_o^2}{R_o^2 - R_i^2} \right) \eta \Omega = C \eta \Omega \qquad (4.7)$$

where C is a constant. The torque generated in the wire is proportional to the angle, θ, through which the wire has been twisted, so $T_i = k\theta$ where k is another instrument constant which can be determined using a liquid of known viscosity. The plot of Ω against θ should be linear with a slope of $C\eta/k$ from which the ratio C/k can be obtained and used to evaluate η from similar plots for other liquids.

The shear rate at different points in the gap between the cylinders is (Exercise 4.2.2) $T_i/2\pi\eta L r^2$ so for a viscometer in which $R_o = 2R_i$ there will be a fourfold variation in shear rate across the gap. Such a device would not be very satisfactory for studying the flow of colloidal suspensions where the viscosity is often a function of shear rate. That would make the above analysis invalid because η would then be a function of the radius. Although it is possible to modify the analysis to take account of a varying viscosity in the gap, the more usual procedure is to keep the gap width, d, as small as possible. The shear rate is then approximately constant and is given by

$$\dot{\gamma} = \Omega R_i/d. \qquad (4.8)$$

Commercial instruments commonly have a radius of several centimetres and a gap width of 1 mm or less (Exercise 4.2.2). Unfortunately, for reasons of convenience, they often choose to rotate the inner cylinder which can give rise to instabilities for liquids of low viscosity like water, even at modest speeds. The flow regime in the bottom of the Couette viscometer is different from that in the annulus between the cylinders and an end-correction may therefore be necessary (Hunter 1989).

Instruments are of two general kinds. Some maintain a known speed of rotation (that is a constant shear rate) and determine the shear stress by some kind of strain gauge. Others apply a constant stress and determine the consequent rotational speed. The latter are usually more sensitive to the behaviour at small deformations. Both types can be electronically automated to varying degrees with results displayed either as a meter reading or a digital readout.

4.2.2 Capillary viscometer

The second type of viscometer, introduced by Ostwald, is very simple in design (Fig. 4.5). It consists essentially of a fine capillary tube connecting two reservoirs. The fluid is drawn up to the top mark in the left-hand side tube in the diagram and allowed to drain out while the tube is held vertical.

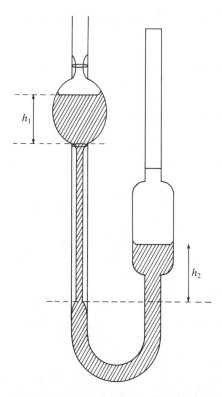

Fig. 4.5. The Ostwald viscometer. The volume of liquid used is such that the height in the right-hand reservoir moves symmetrically about the centre of that reservoir as the height in the left-hand arm falls from the upper to the lower mark. The constriction at the top makes timing more accurate.

The time required for the level to drop to the lower mark is related to the viscosity; the more viscous the liquid, the longer it takes to drain. Comparison is made with a liquid of similar and known viscosity.

To determine the precise form of the relationship between flow rate and viscosity we need to determine how the fluid flows through the tube. The liquid in contact with the wall of the capillary will again be stationary, and successive layers as we move toward the centre of the tube will travel with increasing velocity until a maximum is reached on the axis of the tube. The velocity as a function of distance from the axis of the tube $v(r)$ is called the velocity profile and it can be established by considering the forces on a cylinder of fluid coaxial with the tube and of radius r and thickness Δr (Fig. 4.6).

The driving force for the flow is the difference in pressure at the top and bottom of the capillary and this results in a downward force:

$$\Delta p \times \text{area} = (p_1 - p_2)\, 2\pi r \Delta r. \qquad (4.9)$$

The viscous force which operates on the inside surface, due to the adjoining fluid, also assists the flow since the inner wall is in contact with faster moving fluid so it will be pulled downwards with a force given by

$$\eta \times \text{area} \times \text{shear rate} = (\eta\, 2\pi r\, L\, dv/dr)_{\text{at } r}. \qquad (4.10)$$

Assisting the flow is the mass of the fluid in the thin cylindrical tube itself and the gravitational force is

$$2\pi r \rho g L \Delta r.$$

The flow is opposed by a larger force on the outside surface of the cylinder due to the slower moving liquid further out:

$$(\eta\, 2\pi r\, L\, dv/dr)_{\text{at } r + \Delta r}.$$

Thus we have:

$$\Delta p\, 2\pi r \Delta r - 2\pi r \rho g L \Delta r = \eta\, 2\pi L \left[\left(r\frac{dv}{dr} \right)_{r+\Delta r} - \left(r\frac{dv}{dr} \right)_r \right]$$

$$= \eta\, 2\pi L \left[\frac{d}{dr}\left(r\frac{dv}{dr} \right) \right] \Delta r \qquad (4.11)$$

so that

$$\eta \left[\frac{1}{r}\frac{d}{dr}\, r\left(\frac{dv}{dr} \right) \right] = \left[\frac{p_2 - p_1}{L} - \rho g \right] = G \qquad (4.12)$$

where G is a constant. It is not difficult to integrate this equation to obtain an expression for the velocity (Exercise 4.2.3)

$$v = -\frac{G}{4\eta}(a^2 - r^2) \qquad (4.13)$$

which is the parabolic profile referred to as 'Poiseuille flow' used in the treatment of capillary hydrodynamic fractionation (§3.6). The more useful expression for our present purposes comes from integrating the velocity over the tube cross-section (Exercise 4.2.3) to obtain the flow rate, Q:

$$Q = \int_0^a 2\pi r v\, dr = -\frac{\pi G a^4}{8\eta}. \qquad (4.14)$$

The quantity G is the negative of the pressure gradient down the capillary which drives the fluid, so eqn (4.14) is often written in terms of the volume, V, of fluid flowing through the capillary in time t:

$$V = \frac{\pi a^4 t \Delta P}{8\eta L}. \qquad (4.15)$$

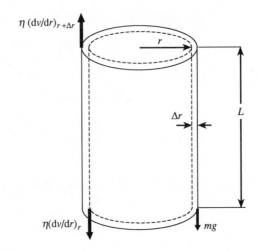

Fig. 4.6. Forces on a cylinder of fluid moving through a cylindrical capillary.

If the liquid levels in the two reservoirs do not change much in height, compared to L, the flow rate will be approximately constant and Q is equal to the volume of liquid between the two markers divided by the drainage time. From measurements of Q the viscosity of the fluid can be determined. More usually a comparison is made with a liquid of known density ρ_0 and viscosity η_0 and the pressure gradient is assumed to be proportional to the density of the fluid. If the drainage times of unknown, t_u and standard, t_0 are compared we then have

$$\frac{\eta_u}{\eta_0} = \frac{t_u \rho_u}{t_0 \rho_0} \tag{4.16}$$

where ρ_u is the density of the unknown.

The Ostwald viscometer is very useful for obtaining accurate viscosities of Newtonian fluids. It is not so satisfactory for colloidal suspensions because in the normal viscometer the shear rate varies widely across the tube diameter. It is necessarily zero on the axis and it can easily be $2000\,\text{s}^{-1}$ at the wall. Correction procedures are available and variants of the Ostwald viscometer are used for measuring colloidal suspensions, especially when they are fairly concentrated so that flow is very slow and only occurs at measurable rates when an external pressure is applied.

Exercises

4.2.1 Integrate eqn (4.5) using the fact that $v = 0$ on the fixed inner cylinder to obtain eqn (4.6).

(*Hint*: Consider the quantity $r(\mathrm{d}\{v/r\}/\mathrm{d}r)$.)

4.2.2 Show that the strain (or shear) rate, $\dot{\gamma}$, in the gap of the Couette viscometer is equal to:

$$\frac{\mathrm{d}\gamma}{\mathrm{d}t} = 2\Omega\left(\frac{1}{r^2}\right)\left(\frac{1}{R_i^2} - \frac{1}{R_o^2}\right)^{-1} = \frac{T_i}{2\pi\eta L r^2}$$

where Ω is the angular velocity of the outer cylinder. Show that for very small gap widths, d, this can be reduced to shear rate, $\dot{\gamma} = \Omega(R_i/d)$. This is the value obtained if the cylinder walls are treated as large flat plates.

4.2.3 Establish eqns (4.13) and (4.14).

4.2.4 Show that $\dot{\gamma}$ at the wall of the capillary viscometer is $a\Delta P/2\eta L$. Estimate the shear rate and the resulting shear stress at the wall of a capillary viscometer of radius 0.5 mm if it takes 100 s to drain a volume of 5 cm³ of water at 15°C.

4.3 Flow of time-independent inelastic fluids

We noted in §4.1 that any system will exhibit elastic (solid-like) behaviour if it is examined on a short enough time-scale, and likewise, will show flow (or liquid-like) behaviour over sufficiently long times. Nevertheless, it is convenient to distinguish those systems which do not show any appreciable elastic behaviour and for which the flow behaviour is not time dependent. Newtonian fluids fall into this category but so too do a lot of colloidal suspensions which are decidedly non-Newtonian, and it is often the departure from Newtonian behaviour which is the important and valuable characteristic of the colloidal system.

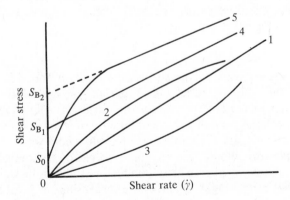

Fig. 4.7. Common behaviour for colloidal suspensions: (1) Newtonian; (2) shear thinning (pseudoplastic); (3) shear thickening (dilatant); (4) ideal (Bingham) plastic; (5) nonideal Bingham plastic. (Type 2 and 5 fluids may or may not exhibit an initial yield stress, S_0.)

Figure 4.7 llustrates some of the possible flow patterns. The simple linear relation between shear stress and shear rate (curve 1) is the Newtonian response of a simple liquid like water at normal shear rates. Sometimes the viscosity decreases as the rate of shear increases (curve 2) and sometimes the reverse occurs (curve 3). There are also many important systems in which the material behaves like an elastic solid for sufficiently small stresses but, once some critical stress is exceeded, the material 'yields', and the resulting flow behaviour may exhibit little if any elastic character. For present purposes we will treat such systems as inelastic, for it is usually the region above the yield value (S_0) which is of most interest. In some such systems the flow, once it begins, is Newtonian (curve 4) but more usually there is a region of varying viscosity which may or may not give way to a Newtonian pattern (curve 5). The viscosity in these cases can be measured in two ways. Strictly, it should always be the ratio of shear stress to shear rate but sometimes we prefer to measure it by the slope of the shear stress versus shear rate curve. That **differential viscosity**, $dS/d\dot{\gamma}$, is often easier to understand in terms of what is happening at the microscopic level of particle–particle interaction.

4.3.1 Stable suspensions of spheres

The simplest colloidal suspension is made up of a collection of smooth, spherical particles immersed in a liquid of the same density, so that they are neutrally buoyant (that is they neither sink nor float). That system was treated by Einstein during his very productive incursion into colloid science at the beginning of this century. He showed that, provided the system was (colloidally) stable and not too concentrated (less than about 10 per cent by volume), one should expect the viscosity of the system to be increased in proportion to the volume of the added spheres:

$$\eta = \eta_0(1 + 2.5\phi) \tag{4.17}$$

where η_0 is the viscosity of the suspending medium and ϕ is the volume fraction of the solid. It turns out then that the size of the spheres is immaterial. They can be all the same or very different. All that matters is the total volume of solid per unit volume of liquid. For all its apparent simplicity this equation is not easy to establish theoretically and it was not easy to test experimentally for colloids. We now have no reason to doubt its accuracy because it has been tested with glass spheres and some biological cells above the colloid size range (Fig. 4.8) and dilute suspensions of spheres appear to approach the value suggested by eqn (4.17) at sufficiently low concentration.

The quantity

$$\frac{\eta/\eta_0 - 1}{\phi} = \frac{\eta_{rel} - 1}{\phi}$$

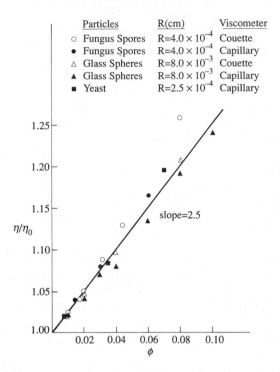

Particles	R(cm)	Viscometer
○ Fungus Spores	$R = 4.0 \times 10^{-4}$	Couette
● Fungus Spores	$R = 4.0 \times 10^{-4}$	Capillary
△ Glass Spheres	$R = 8.0 \times 10^{-3}$	Couette
▲ Glass Spheres	$R = 8.0 \times 10^{-3}$	Capillary
■ Yeast	$R = 2.5 \times 10^{-4}$	Capillary

Fig. 4.8. Verification of Einstein's equation for the viscosity of a suspension of smooth spheres using suspensions of several different sizes. ■: yeast particles, $a = 2.5\,\mu$m; ○, ●: fungal spores, $a = 4.0\,\mu$m; △, ▲: glass spheres, $a = 80\,\mu$m. (Open symbols measured in a Couette and closed symbols in a capillary viscometer.) (Reproduced from Hiemenz (1986) by courtesy of Marcel Dekker Inc.)

is called the intrinsic viscosity, usually written $[\eta]$, and the limiting value of $[\eta]$ as ϕ tends to zero should be 2.5 for spherical particles. Higher values are expected if the particles are nonspherical or if the solid takes up some of the solvent and swells (Exercise 4.3.1). Atkins (1982, p. 825) describes how the intrinsic viscosity is used to estimate the relative molar mass (molecular weight) of a polymer and a more detailed description of its use is given by Hiemenz (1986). Figures 4.9 and 4.10 show how dramatically the behaviour departs from Einstein's relation when the volume fraction rises above 10%.

The form of eqn (4.17) suggests that it is able to account only for the effect of the spheres in isolation (since η depends only on ϕ). As the concentration rises it becomes necessary to take account of the interactions between spheres. Normally that would be done by looking for the coefficient of the next term in the series expansion

$$\eta = \eta_0(1 + 2.5\phi + b\phi^2 + c\phi^3 + \ldots) \tag{4.18}$$

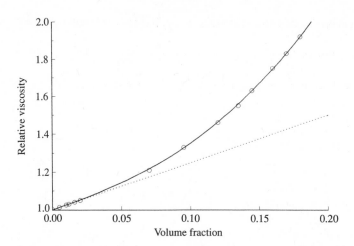

Fig. 4.9. The effect of particle crowding on viscosity. The broken line is drawn according to eqn (4.17), and indicates that these particles show the theoretical intrinsic viscosity of 2.5. The points are experimental and the full curve is drawn using eqn (4.18) with $b = 5$ and $c = 53$. (Jones *et al.* (1991).)

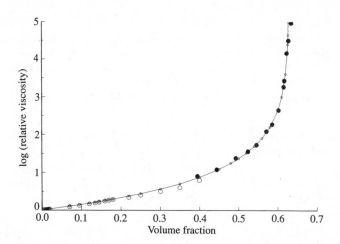

Fig. 4.10. Measured low shear-rate viscosities of concentrated suspensions of the same silica suspension as shown in Fig. 4.9. The curve represents eqn (4.19) with $[\eta]p = 2$ and $p = 0.631$. Note that the relative viscosity is here shown on a logarithmic scale to accommodate values up to 100 000. The intrinsic viscosity used here is $2/0.63 = 3.17$ which is a little higher than the measured value as is indicated by the fact that the curve runs a little above the data points at the lowest volume fractions. (Data from Jones *et al.* (1991).)

and that has been done. After a few false starts in attempting to introduce the effect of sphere–sphere interactions, it is now generally accepted that $b = 6.2$, but the resulting equation only works for a slightly more concentrated suspension than does eqn (4.17). To go to still higher terms in that series would involve more work than could be justified by the result. Fortunately another more empirical approach, introduced by Mooney and developed by Krieger seems to work quite well. It is called the 'particle self-crowding correction' and it yields the relation

$$\frac{\eta}{\eta_0} = \left(1 - \frac{\phi}{p}\right)^{-[\eta]p}$$ (4.19)

where p is an adjustable parameter. This formula can be derived by considering the change in the viscosity $\delta\eta$ caused by an increase in the volume fraction $\delta\phi$. If the suspension before the addition of the extra particles is treated as a Newtonian liquid with a viscosity η, then the Einstein formula suggests

$$\delta\eta = \eta(2.5\,\delta\phi).$$

In order to take account of the fact that the suspension becomes rigid when the particles are close packed, $\delta\phi$ is replaced in this formula by $\delta\phi/(1 - K\phi)$ where $(1 - K\phi)$ is the volume available for the added particles. The idea here is that when $K\phi$ approaches unity, the viscosity change will become very large for any further addition of particles. This is what happens when ϕ is near to the close-pack condition. Then, in differential form

$$d\eta = 2.5\,\eta\,d\phi/(1 - K\phi)$$

or

$$\int_{\eta_0}^{\eta} \frac{d\eta}{\eta} = 2.5 \int_0^{\phi} \frac{d\phi}{1 - K\phi} = -(2.5/K)\left[\ln(1 - K\phi)\right]_0^{\phi}.$$

This can be written (Exercise 4.3.2)

$$\frac{\eta}{\eta_0} = \left(1 - \frac{\phi}{p}\right)^{-2.5p}$$ (4.20)

where $p = 1/K$ is expected to be approximately equal to the volume fraction of the spheres for close packing. In practice, p turns out to be 0.6–0.7. Although this equation is very much better than eqn (4.18) it can be improved still further by replacing the constant 2.5 by the measured value of $[\eta]$ to take account of the particle shape and/or solvation and the possible presence of small numbers of doublets in the suspension.

The problem of applying this treatment to colloids is to find a system which obeys all the requirements of Einstein's assumptions and for which ϕ can be accurately determined. That gets easier the more concentrated the system is, but that raises a new problem. As the concentration rises two things happen: firstly the equation breaks down if the shear rate is raised above a certain value because the suspension becomes non-Newtonian. The viscosity begins to fall slowly until it reaches a limiting value at high shear rate. There are thus two viscosities to be dealt with: one at high shear and the other at low shear with a range in between. Krieger has shown that the high and low shear limits of a stable, concentrated latex suspension obey eqn (4.20) and Jones *et al.* (1991) have shown that the low shear viscosity of a suspension of very small ($r \approx 25$ nm) silica particles can be represented very well, up to almost close pack, when the viscosity has risen to 100 000 times its dilute solution value (Fig. 4.10).

Why does the viscosity decrease with increasing shear rate in this case? That has to do with the relative importance of Brownian motion and the shearing process in determining the motion of the particles. Krieger describes the data by introducing a characteristic shear stress S_i related to the diffusion coefficient of the particles, which increases with their thermal energy and decreases with volume:

$$S_i = kT/\alpha a^3$$

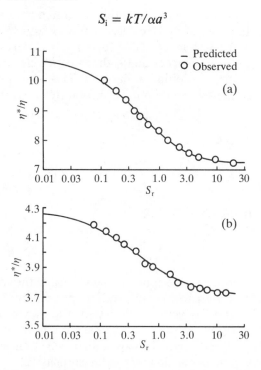

Fig. 4.11. Measured shear viscosities of concentrated latex suspensions as a function of the nondimensional shear stress, $S_r = Sa^3/kT$. The volume fraction is 0.35 for (a) and 0.45 for (b). The curves represent eqn (4.21) with $\alpha = 0.431$.

where α is a dimensionless constant. If S is much less than S_i, Brownian motion is very important but as S rises above S_i the shear regime itself becomes progressively more important. The viscosity of the suspension at various shear rates is then given by

$$\frac{\eta - \eta_\infty}{\eta_0 - \eta_\infty} = \left(1 + \frac{S}{S_i}\right)^{-1} \tag{4.21}$$

where η_∞ and η_0 are the high and low shear limiting viscosities respectively. Figure 4.11 shows how well that equation describes the data. Thus we see that even in the case of perfectly spherical, smooth, stable latex particles the colloidal suspension exhibits non-Newtonian behaviour with a viscosity that depends on both concentration and shear rate.

If the particles are non-spherical, and particularly for rod-shaped particles, the decrease in viscosity with increasing shear rate is even more pronounced. At low shear rates the Brownian motion of the rods makes them rotate and they interfere strongly with one another, so the viscosity is very high. As the shear rate increases, the rods tend to become more aligned with the direction of flow, so that they interfere less with one another and the viscosity decreases. It is not uncommon for suspensions of rod-like particles to show a several-fold change in viscosity rather than the very modest change suggested by Fig. 4.11 for spheres.

The situation might be expected to become much more complicated when the particles are attracted to one another so that they can form some sort of network structure within the fluid. In fact, formulae like (4.19) and (4.21) are found to cover a very wide range of important materials, with some adjustment of the parameters.

4.3.2 Flow of strongly interacting systems

When a liquid containing solid particles is undergoing flow, the presence of the particles has a dramatic effect on the motion of the fluid elements. Even when the particles are far apart they are able to 'see' one another, because the disturbance to the flow, caused by the presence of the particle, falls off very slowly (proportional to r^{-1} where r is the distance from a particle). When we speak of interaction we may be referring simply to this type of **hydrodynamic interaction,** as occurred, for example, in the case of particle self-crowding treated in §4.3.1. There are other situations, however, where we may see an additional effect due to the contact of the particles during a collision. The effects will be particularly obvious if the suspension is to some extent **unstable** in the colloid chemical sense (§2.3) but the flow behaviour will be affected even if the 'bonds' which form between the particles are transient and only exist for a short time during the collision process. It is to these systems which we now turn attention. We will look first at how to describe the flow behaviour and then seek some explanation for the observed behaviour in terms of what we believe is happening at the microscopic (particle–particle interaction) level.

(a) Bingham plastic behaviour   This type of material behaves as a solid for small applied stress ($S < S_B$) and then flows with a constant differential viscosity ($\eta_{PL} = \mathrm{d}S/\mathrm{d}\dot{\gamma}$) for higher shear stresses (curve 4 of Fig. 4.7). S_B is called the **Bingham yield value** and

$$S = S_B + \eta_{PL}\dot{\gamma} \tag{4.22}$$

where η_{PL} is called the **plastic viscosity**. Concentrated slurries of coal dispersed in water† show this type of flow behaviour essentially exactly, as do some food products. It is of more general significance because many engineering designs are based on this expression; the departure of the system from this 'ideal' behaviour is often sufficiently small to be neglected.

It is, however, more usual to observe a nonlinear relation between shear stress and shear rate for stresses above the yield value. The simplest model to fit such systems is the Herschel–Bulkley model:

$$S = S_0 + K\dot{\gamma}^n \tag{4.23}$$

where S_0 is called the **primary yield value**. This behaviour would be detected by a linear plot of $\log(S - S_0)$ against $\log \dot{\gamma}$, showing a slope, n, different from unity. This is, of course, purely an empirical relation which has no underlying theoretical base. It is, therefore, unwise to use it outside the region over which it has been established experimentally.

The physical behaviour of these fluids can be understood in terms of a three-dimensional structure which has sufficient strength to prevent flow if the applied stress is less than S_0. For $S > S_0$ the structure may collapse suddenly to produce flow units which are not subsequently affected by the shear regime. This would give rise to Bingham plastic flow and then $S_0 = S_B$. More usually the breakdown is progressive and the flow units become smaller, and/or more compact, and/or more closely aligned to the streamlines‡ as the shear rate increases; the result in any case is a decrease in differential viscosity ($\mathrm{d}S/\mathrm{d}\dot{\gamma}$) with increasing shear rate.

An intermediate behaviour pattern has been observed in the flow of coagulated dispersions of small spherical particles ($< 1\,\mu$m). Such systems show a nonlinear S versus $\dot{\gamma}$ relation at low shear rates (1–$500\,\mathrm{s}^{-1}$) which

---

† Such suspensions are being studied as a possible alternative to heavy oil as a fuel for electrical power stations.

‡ A streamline in a flowing liquid is the path traced out by a parcel of fluid as the flow progresses.

becomes strictly linear at higher shear rates ($500–3000 \, \text{s}^{-1}$). The intercept on the stress axis from this linear region is again called the Bingham yield value since eqn (4.22) above is obeyed for high values of $\dot{\gamma}$ ($> 500 \, \text{s}^{-1}$) (Fig. 4.7 curve 5).

A substance with a significant Bingham yield value is said to have a high **yield strength**. Yield strength is a very important characteristic of such widely different materials as oil-well drilling muds, paints, toothpaste, paper pulp, crude oil, and many food products, pharmaceutical preparations, and cosmetics. A simple illustration is provided by the behaviour of toothpaste. It is the yield strength which holds the extruded paste as a cylinder. Leave it for a few minutes and the stress produced by gravity is enough to cause it flow down into the bristles of the brush.

(b) Pseudoplastic behaviour   For substances which show a negligible yield value ($S_0 \rightarrow 0$) but a varying differential viscosity, it is often possible to represent the behaviour by a power-law relation:

$$S = K\dot{\gamma}^n \tag{4.24}$$

known as the Ostwald–de Waele model. Obviously the Herschel–Bulkley eqn (4.23) is a simple extension of this expression. The main advantage of eqn (4.24) is its simple two-parameter form and the ease with which it can be differentiated and integrated. Even if it applies only over a limited range of $\dot{\gamma}$ it can be used to extract meaningful rheological data from viscometer flow data provided that it holds for all of the shear rates being experienced by the material. Once the S–$\dot{\gamma}$ data have been acquired they can be fitted to a more elaborate model. Although the flow behaviour is usually represented by the plot of S against $\dot{\gamma}$ (the so-called **basic shear diagram**), it is also possible to represent the behaviour of the material by equations involving the **viscosity** as a function of **shear stress** or as a function of **shear rate**. The choice is one of convenience.

The more parameters the equation contains, the wider is the shear rate range it can be expected to cover. Figure 4.12 compares the two-parameter eqn (4.24) with the three-parameter **Ellis** model:

$$\eta = \frac{\eta_0}{1 + (S/S_i)^n} \tag{4.25}$$

and the four-parameter **Meter model**:

$$\eta = \eta_\infty + \frac{\eta_0 - \eta_\infty}{1 + (S/S_i)^{\alpha - 1}}. \tag{4.26}$$

The Krieger expression (eqn (4.21)) corresponds to putting $\alpha = 2$ in eqn (4.26), which reduces the number of arbitrary parameters to three.

These expressions depend on the fact that the plot of S against $\dot{\gamma}$ is

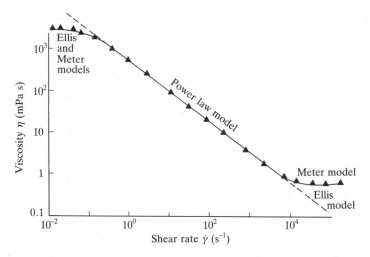

Fig. 4.12. Viscosity as a function of shear rate for a polyacrylic acid solution in water (after Boger). Note that only the Meter model fits the data over the entire range of $\dot{\gamma}$.

usually linear at very low and very high shear rates so that it is possible to define a **zero-shear viscosity**, η_0 and an **infinite-shear viscosity**, η_∞; the apparent viscosity, η, falls smoothly between the two. An alternative representation can be given in terms of the shear rate:

$$\frac{\eta - \eta_\infty}{\eta_0 - \eta_\infty} = [1 + (\lambda\dot{\gamma})^2]^{(n-1)/2} \qquad (4.27)$$

where λ is an empirical (time) parameter. This has some advantages since, in the Couette viscometer in which these systems are usually studied, it is often the shear rate rather than the shear stress which is the quantity which can be most easily varied. We will return to discuss the microscopic origins of these expressions in §4.7.

(c) Dilatant behaviour   The term 'dilatant' refers to the fact that, when a system consisting of irregularly shaped particles of a solid in a liquid is sheared, it increases in volume. It is also observed that the resistance to shear increases (usually fairly dramatically) with increase in shear rate (curve 3 of Fig. 4.17). Thus there is a coincidence of volume expansion and an increase in viscosity with shear rate. There is, however, no necessary connection between the two phenomena and it is possible for each effect to exist in the absence of the other. The increase in viscosity should, therefore, be referred to as **rheological dilatancy** to be more precise. It is most common in systems at very high particle concentration, and in such systems the liquid is effec-

tively acting as a lubricant between the particles. If one attempts to shear too quickly, the particles are pushed more closely together in some regions (though separated in others). The overall effect is to reduce the free movement of the fluid and make the whole system more resistant to shear.

Dilatant behaviour is not as common as pseudoplasticity though when it occurs the consequences can be catastrophic. In a stirring or pumping process, for example, the energy, E, required per unit time per unit volume to keep the flowing system moving is given by the product of the shear stress and the shear rate and for a Newtonian fluid:

$$E = S \times \dot{\gamma} = \eta\dot{\gamma}^2. \tag{4.28}$$

For a dilatant fluid, the relation between S and $\dot{\gamma}$ can usually be represented by a power law (eqn (4.24)) with $n > 1$. The energy dissipation then increases as some power of the shear rate greater than 2. It sometimes happens that, in an industrial process, an attempt to increase pumping or stirring speed moves the liquid into a region where it begins to show dilatancy. The viscosity and the energy input then begin to rise dramatically and the system may either seize up through lack of power or suffer damage from the resulting heating effect. Despite its technological significance, dilatancy has not been studied in the same detail as has pseudoplasticity.

Exercises

4.3.1 The volume fraction, ϕ, in eqn (4.17) is usually calculated from the mass and density of the solid. What is the apparent Einstein coefficient expected to be if the solid spheres absorb some solvent so that their radius swells by 10%?

4.3.2 Establish eqn (4.20).

4.3.3 The following shear stress/shear rate data were obtained for an aqueous solution of methyl cellulose at 18°C (Boger *et al.* 1980). (The first figure is the shear rate (s^{-1}) and the second is shear stress (N m^{-2}).)

0.1400,	0.117;	0.1762,	0.141;	0.2218,	0.169;	0.2793,	0.211;
0.3516,	0.281;	0.4426,	0.352;	0.5572,	0.446;	0.7015,	0.563;
0.8831,	0.687;	1.1117,	0.847;	1.400,	1.076;	1.762,	1.305;
2.218,	1.625;	2.793,	2.010;	3.516,	2.53;	4.426,	3.08;
5.572,	3.79;	7.015,	4.68;	8.831,	5.41;	11.117,	6.53;
14.000,	8.11;	17.620,	9.46;	22.18,	11.49;	27.93;	13.52;
35.16,	16.22;	44.26,	18.92;	55.72,	22.10;	70.15,	26.13;
88.31,	30.00;	111.17,	34.80;	140.00,	40.0.		

See how well these data fit a log–log plot over limited ranges of shear rate and estimate the values of K and n (eqn (4.24)) which best describe them.

4.3.4 Suppose that the particles of a colloidal protein take up water to the extent of m_w grams of water per gram of dry protein. Show that the volume fraction of this wetted colloid ϕ_w is given by:

$$\phi_w = \phi_d \left(1 + \rho_p m_w / \rho_w \right)$$

where ϕ_d is the dry volume fraction. (Assume that the bound water has density ρ_w and ρ_p is the particle density.) If the protein consists of spherical particles, how will its intrinsic viscosity (based on the value of ϕ_{dry}) be affected by this swelling process?

4.3.5 Recast the Ellis model in the form

$$\dot{\gamma} = a \left(S + b S^m \right)$$

and show that it gives a good fit to the data of exercise 4.3.3 over the whole range of $\dot{\gamma}$ using the parameters $\eta_0 = 0.794\,\text{N m}^{-2}\,\text{s}$, $S_i = 21.554\,\text{N m}^{-2}$, and $m = 2.027$. Compare this with the fit using eqn (4.21) with $\eta_\infty = 0$. (That is the low shear Meter model with $\alpha = 2$.)

4.4 Time-dependent inelastic fluids

It sometimes happens that the apparent viscosity of a suspension depends both on the shear rate and the time for which the system has been sheared. This is an indication of the fact that the relaxation time† for the material is of the same order as the time over which the system is being studied. In other words, the 'structure' in the system, which determines the variable shear viscosity, is being altered at a rate which is observable over the time period of the measurement. It is always possible to encounter this sort of phenomenon if one chooses the right time-scale, but it is sometimes impossible to avoid. If the system has a short relaxation time (say $< 10^{-3}\,\text{s}$) it will be possible to construct a curve like Fig. 4.12 with little or no evidence of time dependence (except perhaps at the very highest shear rates). But when the relaxation time is of the order $10\text{--}10^3\,\text{s}$ or even higher (as it can well be with suspensions of highly anisometric particles) it becomes much more difficult to obtain meaningful data.

The most common phenomenon of this type is called **thixotropy** where the apparent or differential viscosity at a particular shear rate decreases with time. Less frequently, the opposite behaviour, called **negative thixotropy**, occurs. In thixotropic materials, shearing at a given shear rate produces a gradual breakdown in the 'structure' with time. If it is allowed to proceed for long enough one can obtain an 'equilibrium' or steady-state curve which is usually pseudoplastic in nature. More commonly the time constraints make

---

† When a system at equilibrium is subjected to a sudden change in the external conditions it will take time to adjust itself to the new conditions; the characteristic time for that process is called the relaxation time for the system.

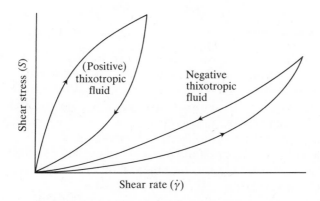

Fig. 4.13. Hysteresis loops for thixotropic fluids (both positive and negative).

it impossible (or inconvenient) to wait for the steady-state behaviour to be established.

Thixotropy can be detected by taking the system from low to high shear rate and back again. If the curves in each direction are different one obtains a **hysteresis loop** (Fig. 4.13). The area inside the loop depends on the past shear (and possibly thermal) history of the sample. The more time that is spent at each data point, the more chance the system has to reach its steady-state behaviour and so the smaller is the area of the loop.

A related behaviour occurs in some systems at low shear rates where the rate at which the structure can be rebuilt is increased by gentle shearing. This is referred to as **rheopexy**. It can be recognized by the fact that the system recovers some of its pre-sheared viscosity at a somewhat faster rate when it is sheared slowly compared to when it is left to stand. Note that it is not the same as negative thixotropy though the two can be confused if only a limited number of measurements is made at one (low) shear rate.

Figure 4.13 shows the expected behaviour of a negatively thixotropic material. This phenomenon is not common in particulate systems. It occurs more readily in polymer solutions, however, where rapid shearing can induce considerable entanglement. Such systems will usually relax back to a lower viscosity when left undisturbed. The thermal motion of the polymer chains causes them to disentangle and so the shear-induced 'structure' gradually breaks down.

Thixotropy has been observed in crude oils, in bentonite (mont-morillonite) clays (§1.4.4), and in some food, pharmaceutical and cosmetic products. It is becoming increasingly important in paints. In engineering situations (like pipe-line flow) the thixotropy can often be ignored since its effects become less important if the system is undergoing continuous shear. It can, however, be important in the start-up phase after a flow line has been shut down for some time.

4.5 Viscoelastic fluids

Fluids which have some elastic character can exhibit quite a number of
unusual, even bizarre, flow effects. Indeed, in several respects their behaviour is the opposite of purely viscous materials. When, for example, a
viscoelastic material is stirred with a rod (Fig. 4.14) it moves towards the rod
and begins to climb up it, in contrast to a viscous liquid which forms a
vortex. The same behaviour (called the 'Weissenberg effect') is observed in
Couette flow: the viscoelastic fluid tends to accumulate near the inner wall
instead of being thrown towards the outer cylinder, as occurs for the purely
viscous fluid at sufficiently high rotational speeds.

When a stream of viscous fluid emerges from a vertical pipe, its diameter
tends to be smaller than that of the pipe, whereas a viscoelastic fluid in the
same circumstances has a larger diameter than the pipe (Fig. 4.14(c). (This
phenomenon is known as **die swell** since it is often encountered when
polymer solutions or melts or colloidal suspensions or pastes are extruded
from a die.)

Some other flow effects are indicated in Fig. 4.14. The strangest is the
'tubeless syphon' effect, in which a syphon continues to function even when
the upper end is removed some little distance from the liquid (Fig. 4.14(f)).
One can observe such elastic behaviour when pouring hair shampoo and
other surfactant systems, but some high molar mass polymer solutions
exhibit these effects to an extraordinary degree.†

4.6 Measurement of non-Newtonian flow characteristics

4.6.1 Time-independent inelastic fluids in Couette flow

We return now to the question of how to obtain meaningful rheological data
from flow situations when the fluid is inelastic but non-Newtonian. In
general the shearing stress and rate of shear will depend upon the position
in the viscometer but the simplest procedure is to use a viscometer with a
narrow gap width so that the shear rate remains essentially constant across
the gap (Exercise 4.2.2).

Unfortunately, there is a limit to which the gap can be reduced. To treat
the fluid between the cylinders as a homogeneous material it is necessary to
have a gap which is much larger (at least 100 times) than the size of the largest
particles or aggregates in the suspension. To satisfy this requirement, most
commercial viscometers tend to have rather large gaps so that the ratio
R_o/R_i departs significantly from unity. Careful analysis of the data then
becomes essential.

---

† Cheese fondue (molten cheese) is a good example of a viscoelastic fluid.

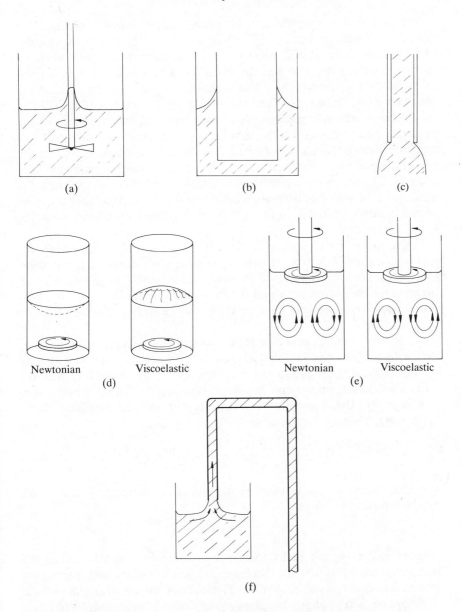

Fig. 4.14. Flow behaviour of viscoelastic fluids. (After Bird and Curtiss (1984).) (a) The Weissenberg effect; (b) in Couette flow; (c) die swell; (d) and (e) influence of a rotating disc on the flow pattern, and (f) the open syphon.

It is important to understand that the **distribution of stress** in the gap does not depend on the properties of the fluid. (The flow regime in each cylinder of fluid in the annulus will adjust itself so that the torque is constant from the inner to the outer cylinder; the stress on each cylinder of fluid is fixed by its position in the annulus.) The experimental data consists of measurements of applied torque as a function of rotational speed (see §4.2.1). When the system is undergoing steady shear, the torque, T, on a cylinder of fluid of radius r and length L, in the annulus between inner and outer cylinders and coaxial with them is $T = 2\pi r^2 LS$ and this torque is constant across the gap and equal to the external torque, T_i or T_o, on the stationary cylinder (§4.2.1). The determination of the shear rate is, however, a little more difficult since the velocity distribution across the gap depends on the properties of the fluid. Fortunately, one can determine those properties from the experimental data.

Equation (4.3) still holds for the rate of strain of an inelastic non-Newtonian fluid. One way to analyse the data is to assume that this shear rate is some arbitrary function of the stress and so to obtain a general relation between the rate of rotation and the measured torque. Details of the procedure are given elsewhere (Hunter 1989) and we will not pursue it further here.

A less general but quite adequate method in many cases is to assume a particular (empirical) relationship between S and $\dot{\gamma}$ for the fluid and analyse the flow regime in the gap to arrive at the correct way in which to treat the data. The best known example of this procedure is the Reiner–Riwlin equation based on the assumption that the fluid is an ideal Bingham plastic (§4.3.2(a)). They derive (Exercise 4.6.1)

$$\Omega = \frac{T_i}{4\pi L\eta_{PL}}\left(\frac{1}{R_i^2} - \frac{1}{R_o^2}\right) - \left(\frac{S_B}{\eta_{PL}}\right)\ln\frac{R_o}{R_i} \qquad (4.29)$$

provided that all of the material in the annulus is undergoing shear. This expression can be written

$$\Omega = \eta_{PL}^{-1}\left(k_1\theta + k_2\right)$$

where θ is the angular deflection of the torsion wire and k_1 and k_2 are instrument constants with the first being independent of the material and the second depending on the yield value. The slope of the plot of Ω against θ can therefore be used to obtain the plastic viscosity and the intercept will give a measure of the yield value.

4.6.2 *Some experimental considerations*

If the gap width in the Couette viscometer is fairly large, the stress in the gap may not be large enough to cause flow of a Bingham fluid right across the annulus, especially at low shear rates. There will develop a region near one of the walls in which the material moves as a rigid body with the same velocity as the wall so that the effective gap width is reduced. The problem, once recognized, can be overcome, and one can also use the other models described above (eqns (4.25) and (4.26)) to derive reliable rheological data on non-Newtonian fluids (Hunter 1989).

In the derivation of the above equations it is assumed that the fluid in contact with the inner and outer cylinders moves with the same velocity as the adjoining solid. This is the expected behaviour for simple fluids, but for a colloidal dispersion this is not necessarily so. If the particles are larger in size than the roughness of the surfaces, it is possible for a phase separation to occur so that the surface is covered with a thin layer of suspension medium which may act as a lubricant, allowing the suspension to slip with respect to the solid surface. The problem can be alleviated by deliberately roughening the cylinder surfaces or, in extreme cases, introducing knurls, ridges, or even spikes.

The viscosity of a molecular fluid normally decreases exponentially with temperature so it is important to control the temperature in any viscometry. Not only is it essential to protect the suspension from heat gain or loss due to ambient temperature fluctuations but it is also necessary to make provision for the heat generated by the flow process itself, especially at high shear rates. Obviously, a narrow gap is again an advantage, and constructing the outer cylinder of metal with a surrounding thermostatted jacket is a common procedure.

4.6.3 *Measurements in the capillary viscometer*

We discussed the behaviour of a Newtonian fluid in a capillary (Ostwald) viscometer in §4.2.2. For a non-Newtonian fluid it is easier to pursue the analysis by way of the stresses rather than the velocity, just as it was in the Couette. For a Newtonian fluid the shear stress at the wall is (Exercise 4.2.4):

$$S_w = \eta \times \dot{\gamma}_{wall} = a\Delta p/2L \qquad (4.30)$$

where a is the capillary radius and L is its length. The same is true for any inelastic or viscoelastic fluid, and for fluids having a yield strength. The stress throughout the capillary is also the same as for a Newtonian fluid. It is equal to $S_w r/a$ at different radii, so the stress rises linearly from the axis, where it is zero, to a maximum value at the wall. The relation between wall shear stress and pressure drop (eqn (4.30)) allows us to develop a relation between volume flow rate and the pressure across the capillary for a more general fluid.

(a) Flow rate versus pressure drop   The volume flow rate, Q, in the z direction through the capillary, is related to the fluid velocity, v_z, by eqn (4.14):

$$dQ = v_z 2\pi r \, dr$$

which can be integrated by parts to give

$$Q = \pi \left[v_z r^2 - \int r^2 dv_z \right]_0^a = -\pi \int_0^a r^2 dv_z \qquad (4.31)$$

assuming there is no slip at the tube wall so that $v_z = 0$ for $r = a$.

The shear rate for fully developed (laminar) flow in a tube is in this case $(-dv_z/dr)$ and this will be a function of the shear stress:

$$-dv_z/dr = f(S(r)). \qquad (4.32)$$

(The negative sign is introduced because v_z decreases from its maximum (on the axis) as r increases.) Combination of eqns (4.31) and (4.32) then leads to (Exercise 4.6.4)

$$\frac{Q}{\pi a^3} = \frac{1}{S_w^3} \int_0^{S_w} S^2 f(S) \, dS. \qquad (4.33)$$

Just as in the case of the Couette viscometer (§4.6.1) this expression can be used in several ways – it can be integrated for a particular fluid model to obtain the volumetric flow rate–pressure drop relationship for flow through a circular pipe or it can be differentiated to obtain an expression for shear rate at the wall, independent of the fluid model.

If, for example, we choose a power-law model for the fluid:

$$S(r) = K(-dv_z/dr)^n \qquad (4.34)$$

so that $f(S(r)) = -dv_z/dr = (S(r)/K)^{1/n}$ then, it is not difficult to show that (Exercise 4.6.6)

$$\frac{Q}{\pi a^3} = \frac{n}{3n + 1} \left(\frac{a}{2K} \frac{dp}{dz} \right)^{1/n} \qquad (4.35)$$

where $dp/dz = -\Delta p/L$ is the pressure drop over the length of the capillary. A fluid obeying eqn (4.35) would give a linear plot of volume flow rate against the logarithm of the pressure drop, with a slope of $1/n$. Knowing a one could then determine K and hence the relation between shear stress and shear rate for that fluid. Corresponding expressions can be written for other fluids (Hunter 1989).

It turns out that, for a power-law fluid (especially if n is small), the pressure drop is much less sensitive to flow rate than it is for a Newtonian fluid. The flow profile for such fluids is very different from that shown by

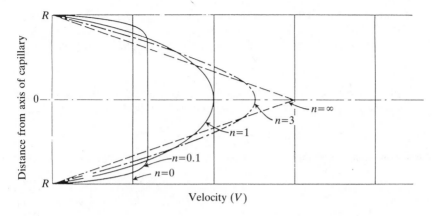

Fig. 4.15. Velocity profiles for power law fluids, corresponding to the same average velocity. (After van Waser *et al.* (1963).)

a Newtonian fluid (Fig. 4.15). As n decreases below unity the velocity becomes constant across a large part of the tube (a phenomenon called 'plug flow'). For dilatant systems ($n > 1$) the profile is sharper than the parabolic result obtained for a Newtonian fluid. There are other more general procedures which can be applied to data from a capillary viscometer but space does not permit us to examine them here (see Hunter 1989).

Capillary viscometers may be constructed of glass with the flow occurring solely as a consequence of a difference in hydrostatic head between inlet and outlet (§4.2.2). For concentrated colloidal suspensions or polymer solutions, however, it is usually necessary to employ gas pressure to drive the fluid through. There are various problems associated with this procedure and a number of significant sources of error which are discussed in more specialized texts (van Waser *et al.* 1963). Some of these errors can be estimated and corrected for directly whilst some can be minimized by proper design. We will not examine these questions in any further detail.

Exercise

4.6.1 For an ideal Bingham fluid one can write shear rate $= r \, d\omega/dr = (S - S_B)/\eta_{PL}$. Use this to derive the Reiner–Riwlin equation (4.29) assuming that the fluid is moving with angular velocity Ω at the outer cylinder ($r = R_o$) and zero *only* at the inner cylinder ($r = R_i$).

Hint: Refer to the derivation of eqn (4.7) and show first that

$$T_i = 2\pi r^2 L \left[\eta_{PL} \left(\frac{dv}{dr} - \frac{v}{r} \right) + S_B \right].$$

4.6.2 Show that if the Bingham yield value S_B is such that only some of the fluid is moving ($S_o < S_B < S_i$ where subscripts o and i refer to outer and inner cylinders) then the critical radius beyond which the fluid is stationary is

$$R_{crit} = R_i (S_i/S_B)^{1/2}.$$

Hence show that the critical stress S_{crit} (measured at the bob) which must be exceeded so that flow occurs throughout the annulus is given by

$$S_{crit} = S_B (R_o/R_i)^2.$$

4.6.3 Show that the Reiner–Riwlin equation can be written in the form:

$$\text{shear rate} = \dot{\gamma} = \frac{(S_i - S_B)\Omega}{T_i k_1 - S_B k_2}$$

where k_1 and k_2 are apparatus constants. The data in the linear regime (where $T_i \propto \Omega$) is treated by plotting this function $\dot{\gamma}$ against $S_i (= T_i/2\pi R_i^2 L)$ and determining the slope (which is η_{PL}).

4.6.4 Establish eqn (4.33), using the fact that $S = S_w r/a$ where S_w is the shear stress at the wall.

4.6.5 Show that for a Newtonian fluid in an Ostwald viscometer $4Q/\pi a^3 = 8V/D = S_w/\eta$, where V is the average fluid velocity ($=Q/\pi a^2$) and D is the tube diameter. (The quantity $8V/D$ appears as a natural parameter for the other flow models. It is common practice to plot this function (or its logarithm) against the pressure drop in order to represent the flow behaviour.)

4.6.6 Derive eqn (4.35). What is the effect of increasing the velocity fourfold when the flow behaviour index (n) is 0.3? Compare this with a fourfold increase in velocity for a Newtonian fluid.

4.7 Microscopic description of flow behaviour

We must now examine some of the physicochemical descriptions which have been proposed to underpin the mathematical models of flow behaviour described in §4.3. Most discussions of pseudoplastic or plastic behaviour assume that the decrease in viscosity with increasing shear rate is due to a gradual reduction in the amount of 'structure' in the system. They differ mainly in the way in which the applied stress is partitioned between the structural (breakdown) effects and the viscous effects. The fact that it is necessary to use a relationship like eqn (4.21) even for hard (that is noninteracting) spheres tells us that the 'structure' need not correspond to actual bond formation in the normal sense. The variation in viscosity in the hard-sphere systems is, however, quite small — of the order of 15–30 per cent say — whereas there are many systems in which the viscosity changes over several orders of magnitude between its zero shear and high shear limits. It is these that we must now consider.

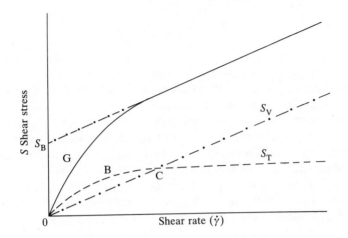

Fig. 4.16. Partitioning the shear stress between viscous flow and structural break-down. On the curvilinear part of the S–$\dot{\gamma}$ relation at, say G, the stress, S_v, required to support viscous flow is subtracted from the total stress to obtain the broken line 0BC which therefore represents the stress S_T involved in structure breakdown as a function of $\dot{\gamma}$.

4.7.1 Kinetic interpretation of non-Newtonian flow

Williamson (1929) made the first quantitative attempt to describe shear thinning behaviour in terms of structural breakdown. He separated the applied shear stress into two parts: one part was assumed to be responsible for the breakdown process and the other part maintained the viscous flow. His method is most appropriate for systems for which the S–$\dot{\gamma}$ curve becomes strictly linear above some critical shear rate (say $\dot{\gamma}_c$) and has been used (Ekdawi and Hunter 1983) for describing the flow behaviour of coagulated sols at low shear rate. The energy dissipated per unit volume per unit time is equal to the product $S\dot{\gamma}$ and this is the basis of the partitioning procedure which is illustrated in Fig. 4.16. The stress required to support the viscous flow S_v is assumed to be given by the Newtonian expression

$$S_v = \eta_{PL}\dot{\gamma} \tag{4.36}$$

for all values of $\dot{\gamma}$ and this is represented by the line 0C. The total stress is then

$$S = S_T(\dot{\gamma}) + \eta_{PL}\dot{\gamma}. \tag{4.37}$$

This can be written

$$S = \frac{S_T(\infty)\dot{\gamma}}{C + \dot{\gamma}} + \eta_{PL}\dot{\gamma} \qquad (4.38)$$

where $S_T(\infty)$ is the value of S_T as $\dot{\gamma} \to \infty$ and C is a measure of the curvature of the S–$\dot{\gamma}$ function at low shear rates. For $C = 0$ the expression obviously simplifies to the Bingham relation (eqn (4.22)). For $S_T = 0$ it represents simple Newtonian behaviour.

The next step is to find some rationale for the function $C(\dot{\gamma})$ and this is done by examining the rate of 'link' formation and breakdown during shear. A typical approach (Cross 1965) assumes that the suspension consists of chains of particles with an average of L links per chain. The links are formed by Brownian motion with a rate constant k_2 and are ruptured by both Brownian motion (rate constant k_0) and by shear (with rate constant $k_1\dot{\gamma}^n$. The rate of change in the number of links is then given by

$$\frac{dL}{dt} = k_2 P - (k_0 + k_1\dot{\gamma}^n)L \qquad (4.39)$$

where P is the number of particles. (Note that build-up of structure by the shearing process (rheopexy) is denied in this approach.) In the steady state $dL/dt = 0$, and so $L = k_2 P/(k_0 + k_1\dot{\gamma}^n)$. If $L = L_0$ when $\dot{\gamma} = 0$, then $L_0 = k_2 P/k_0$ and so

$$\frac{L}{L_0} = \frac{k_0}{k_0 + k_1\dot{\gamma}^n} = \frac{1}{1 + \kappa\dot{\gamma}^n} \qquad (4.40)$$

where $\kappa = k_1/k_0$. The increase in viscosity as $\dot{\gamma}$ decreases is assumed to be proportional to the number of links:

$$\eta = \eta_\infty + BL \qquad (4.41)$$

where B is a constant. Then putting $\eta = \eta_0$ when $L = L_0$ gives $\eta_0 - \eta_\infty = BL_0$ and so

$$\frac{\eta - \eta_\infty}{\eta_0 - \eta_\infty} = \frac{L}{L_0} = \frac{1}{1 + \kappa\dot{\gamma}^n}. \qquad (4.42)$$

Although expressed in terms of the shear rate rather than the shear stress this is clearly equivalent to eqn (4.21) or (4.26). The only problems with this approach are that there is insufficient theoretical basis for one to be able to estimate what value the shear rate dependence parameter (n) should have and there is no way in which one can introduce the energy involved in breaking links. Nor is it possible to incorporate any knowledge of the actual structure of the suspension into the analysis. All suspensions are assumed to consist of chains of particles (rather like linear polymers). It is clear from Fig. 4.12 however, that this type of equation can represent the viscosity behaviour of some pseudoplastic systems over a very wide range of shear rates; it has also been found to represent a very wide range of systems over a rather more limited range of shear rates. Other possible procedures produce similar results (Hunter 1989) but we will not pursue them further here.

This type of approach can be extended in a way which does allow the flow behaviour to be related directly to the properties of the colloid, at least for coagulated sols containing spherical particles less than $1 \mu m$ in size (Hunter 1989).

4.7.2 Time-dependent systems: kinetic interpretation of thixotropy

Denny and Brodkey (1962) applied a reaction kinetics approach to flow behaviour and obtained a result similar to Krieger's equation (4.21). They set up the simple reaction scheme

$$\text{unbroken bonds} \leftrightarrow \text{broken bonds} \qquad (4.43)$$

with forward and reverse rate constants k_1 and k_2 to represent the structural effects. They write for the rate of structural breakdown

$$-\frac{\mathrm{d}(\text{unbroken})}{\mathrm{d}t} = k_1' (\text{unbroken})^n - k_2 (\text{broken})^m$$

and so:

$$-\mathrm{d}\left(\frac{\eta - \eta_\infty}{\eta_0 - \eta_\infty}\right) \bigg/ \mathrm{d}t = k_1' \left(\frac{\eta - \eta_\infty}{\eta_0 - \eta_\infty}\right)^n - k_2 \left(\frac{\eta_0 - \eta}{\eta_0 - \eta_\infty}\right)^m. \qquad (4.44)$$

(Thus the viscosity is introduced by assuming that it is a linear function of the amount of 'unbroken structure' which is a maximum at η_0 and a minimum at η_∞.) The rate constant k_2 is assumed to be independent of shear (that is only Brownian motion leads to restructuring) whilst the breakdown rate constant depends on shear rate:

$$k_1' = k_1 \dot{\gamma}^p$$

where p is a constant which reflects the 'shear sensitivity' of the material. Under steady-state conditions $\mathrm{d}\eta/\mathrm{d}t = 0$, and for $n = m = 1$ eqn (4.26) is generated. The novelty of this approach lies in the solution for nonsteady-state conditions. Choosing possible values for m and n (for example, $m = 2$ and $n = 1$) they obtain analytical solutions for the integrated form of eqn (4.44). Given the values of the parameters m, n, $\dot{\gamma}$, η_0, η_∞, and p derived from steady-state measurements it is then possible to evaluate k_1 and k_2 for a thixotropic material. For the heavy thixotropic mineral oil used in their studies they found $k_1 = 1.21 \times 10^{-14}$ and $k_2 = 0.94 \times 10^{-3} \mathrm{s}^{-1}$ so that the 'equilibrium constant' for eqn (4.43) is $K = k_1/k_2 = 1.29 \times 10^{-11}$. The very low value of k_1 indicates why the material is thixotropic: the rate of breakdown is so slow that it requires a considerable time to establish the steady-state structure at a given shear rate. They also found good agreement between the estimate of K from the time-dependent (thixotropic) data and the value obtained in the steady-state (pseudoplastic) regime (0.89×10^{-11}). Unfortunately, the analysis is rather lengthy and requires a large amount of good data to enable reasonable values of the parameters to be extracted. Its importance lies in the demonstration of the link between thixotropy and pseudoplasticity.

If the time for the establishment of the steady-state structure at a given shear rate is long compared with the measuring time, then the system will exhibit thixotropy. Pseudoplastic behaviour then appears as the limiting form of thixotropy when the time between successive measurements is long compared with the relaxation time for the structure in the system, which can then always exhibit its steady-state behaviour.

Exercises

4.7.1 Show that the time taken for a particle to undergo Brownian diffusion through a distance equal to its radius is about $3\pi\eta_0 a^3/kT$. (Refer to Chapter 2.)

References

Atkins, P. W. (1982). *Physical chemistry*, (2nd edn), p. 825. Oxford University Press.

Bird, R. B. and Curtiss, C. F. (1984). *Physics Today*, **37**, 36–43.

Boger, D. V., Tiu, C., and Uhlherr, P. H. T. (1980). *Introduction to the flow properties of polymers*. R.A.C.I. and British Society of Rheology, Australia.

Cross, M. M. (1965). *Journal of Colloid and Interface Science*, **20**, 417.

Denny, D. A. and Brodkey, R. S. (1962). *Journal of Applied Physics*, **33**, 2269–74.

Ekdawi, N. and Hunter, R. J. (1983). *Journal of Colloid and Interface Science*, **94**, 355–61.

Hiemenz, P. C. (1986). *Principles of colloid and surface chemistry*, (2nd edn), pp. 198–207.

Hunter, R. J. (1989). *Foundations of colloid science*, Vol. II, Ch. 18. Oxford University Press.

Jones, D. A. R., Leary, B., and Boger, D. V. (1991). *Journal of Colloid and Interface Science*, **147**, 479–95.

van Waser, J. R., Lyons, J. W., Kim, K. Y., and Colwell, R. E. (1963). *Viscosity and flow measurement*. Interscience, New York.

Williamson, R. V. (1929). *Industrial & Engineering Chemistry*, **21**, 1108.

5

THERMODYNAMICS OF SURFACES

5.1 Introduction

Although some aspects of colloidal behaviour are far removed from equilibrium situations, there are many which reach an equilibrium state very quickly and for which the methods of equilibrium thermodynamics are very well suited. Indeed surface, or **interfacial**, chemistry is one of the most important areas of application of classical thermodynamics because thermodynamics allows us to extract much useful information from a few seemingly unrelated observations.

We begin with a discussion of the important concepts of surface tension and surface energy which give rise to a difference in pressure across a curved interface. Because small particles necessarily have highly curved surfaces, that pressure difference has important effects on colloidal particles. We then examine a number of phenomena connected with the way a liquid sticks to a surface: the wetting of a solid by a liquid has important implications in many fields including agriculture, mineral ore preparation, and detergency.

5.2 Surface energy and its consequences

5.2.1 *Surface tension and surface free energy*

The term **surface tension** refers to the observation that the surface of a liquid behaves as though it were covered with a thin 'skin' which resists (to some extent) attempts to puncture it. Close observations of insects walking on a water surface show how the surface bends under the weight of each foot, and a needle, (especially if it is a bit greasy) when placed gently on a water surface will rest there. The existence of surface tension can be expected from the difference in energies between molecules at the surface and molecules in the bulk phase of a material.

Consider first a homogeneous liquid or solid, consisting of molecules of type A, in equilibrium with its vapour. Suppose that $v_{AA}(r)$ is the potential energy of interaction between two molecules of type A separated by a distance r, when the potential energy of an isolated molecule is taken as zero. Assuming that nearest-neighbour interactions are dominant in a condensed phase, and that potential energies of interaction are pair-wise additive, the energy $E_{A,bulk}$ *per molecule* in the bulk phase is:

$$E_{A,bulk} \approx \tfrac{1}{2} z_{AA,bulk} v_{AA}(r_b) \tag{5.1}$$

where $z_{AA,bulk}$ is the number of molecules in the shell of nearest neighbours in the bulk phase and r_b is the average distance of these molecules from the central molecule.

The energy $E_{A,S}$ per molecule at the surface is, similarly

$$E_{A,S} \approx \tfrac{1}{2} z_{AA,S} v_{AA}(r_S) \tag{5.2}$$

where we expect $r_s \approx r_b$ and $z_{AA,S} \approx \tfrac{1}{2} z_{AA,bulk}$. Remembering that $v_{AA}(r)$ is negative, it is clear that there is an increase in potential energy when a molecule is taken from the bulk and placed in the surface (that is, *work must be done to create any new surface*).

If the interface is between two liquids, with molecules A and B, then a molecule of A will lose about half of its interaction with other A molecule but gain a similar number of interactions with B molecules. The same is true for the B molecules. The net change in energy, δE, is given by

$$\delta E = (E_{A,S} - E_{A,\text{bulk}}) + (E_{B,S} - E_{B,\text{bulk}}) \tag{5.3}$$

and if this is positive the interface will tend to shrink to a minimum possible area. However, if δE is negative, the interface will tend to grow and the phases will tend to dissolve in one another.†

The number of molecules in the surface is usually only a small proportion of those in the bulk. For example, in a spherical droplet of water of volume 1 cm³, only about one in five million of its molecules are in the surface. The energy of the surface molecules will make an important contribution only for: (i) processes where the bulk energy is unaffected or (ii) systems that are so subdivided that the surface energies are, in any case, comparable to bulk energies. Case (ii) does not apply to most lyophobic colloids though it does apply to lyophilic ones (§1.3). For most of the processes we are concerned with, the surface energy is important because the bulk energy does not change significantly during the process.

The work, δw, required to create new surface is proportional to the number of molecules brought from the bulk to the surface, and hence, to the area δA, of the new surface, so that $w \propto \delta A$ or

$$\delta w = \gamma \, \delta A \tag{5.4}$$

where γ, the constant of proportionality, is defined as the surface energy or the specific surface free energy. Note that it has dimensions of **force per unit length** and for a pure liquid it is numerically equal to the **surface tension**.

To see the relation between the two concepts, consider the following thought experiment. If an arbitrary surface is extended as in Fig. 5.1, the increase in area is given by $\delta A = l\delta x$. The work, δw, done in increasing the area is

$$\delta w = \gamma \delta A = \gamma l \delta x. \tag{5.5}$$

If that work is done by a force F, which is applied to the perimeter, then $\delta w = F\delta x$ and so, comparing this with eqn (5.5), $\gamma = F/l$. That is γ acts like a force per unit length of the perimeter opposing any attempt to increase the area, i.e. like a restoring force or tension (§4.1.1). γ is normally measured in millinewtons per metre (mN m⁻¹), equivalent to the c.g.s unit of dyne cm⁻¹). For a liquid–liquid or a liquid–vapour interface, the equilibrium value of γ is independent of direction, so that the surface is in a state of uniform tension in every direction, if the surface is quiescent. Typical values

---

† Strictly we should deal with the free energy, which includes an entropy contribution; that would favour dissolution even if the energy were slightly positive.

Thermodynamics of surfaces

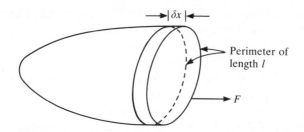

Fig. 5.1. Increasing the area of a surface of arbitrary shape.

of γ for some pure liquids and interfacial tensions for liquid–liquid systems are given in Table 5.1.

An important aspect of the above argument about 'extending the surface' or 'creating fresh surface' is that we mean new surface with the *same properties* as the original surface. In other words, the surface is created by adding new molecules from the bulk, maintaining the same properties and with the molecules in their equilibrium configuration. If the increase in surface area were to be achieved by stretching the average distance between molecules (in the surface and in the bulk) then the extension of the surface would be accompanied by an increase in the bulk energy of the system. That would correspond to an elastic deformation (§4.1.1) and would require a lot more energy. The surfaces of solids also possess a tension so there is an extra free energy due to the surface of the solid. It is, however, difficult to treat, either experimentally or theoretically, in the same way as a liquid surface. The basic problem is that whether the surface is freshly created (by cleaving a crystal) or has been aged for a long time, it will still not be in equilibrium; the atoms are not usually free to move to positions of lowest energy. In a liquid, on the

Table 5.1 Surface and interfacial tensions of some liquids (in mN m^{-1}) at 293 K[4] (Aveyard and Haydon 1973)

	Liquid–vapour	Water–liquid	$-d\gamma/dT$ (liq.–vap.) (mN m^{-1} K^{-1})
Water	72.75	–	0.16
Octane	21.69	51.68	0.095
Dodecane	25.44	52.90	0.088
Hexadecane	27.46	53.77	0.085
Benzene	28.88	35.00	0.13
CCl$_4$	26.77	45.0	–
Mercury	476	375	–

$d\gamma/dT$ for the hydrocarbon–water interface is 0.09 mN m^{-1} K^{-1}.

other hand, the atoms or molecules take up their equilibrium positions in a fraction of a second.

5.2.2   Pressure differences across curved surfaces: the Young–Laplace equation

The tension in a surface must be balanced by an equal and opposite force. That is most obvious when one considers a soap bubble, consisting as it does of a gas inside a thin film of liquid. That thin film consists of many layers of molecules, and the innermost part of the film, away from both surfaces is, therefore, like bulk liquid. Shrinking the bubble would allow more molecules to go into the innermost (bulk) part of the liquid. Why then does it not shrink to reduce the total surface area? The answer is that it does: it shrinks until the pressure builds up to a value high enough to prevent further shrinking. There is, therefore, an excess pressure inside the bubble compared to that outside, and the smaller the bubble is, the larger is that excess pressure. That fact can easily be verified using the apparatus shown in Fig. 5.2.

The derivation of a formula for the excess pressure inside such a bubble is not difficult. Consider a spherical bubble of radius r. The pressure outside the bubble is p' and inside the bubble is p''. Equilibrium is established when the energy gained by shrinking the size of the bubble is just equal to the work done in compressing the air inside. The energy gained by shrinkage $= \gamma\,\delta A$ where A is the area of *both the inside and the outside* of the bubble film

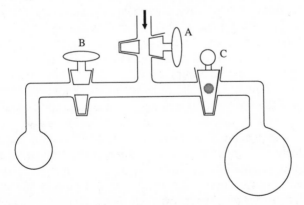

Fig. 5.2.   The apparatus has stopcocks which allow one to blow bubbles of different radius on the two arms whilst keeping them isolated from each other. When they are connected by opening stopcocks B and C, it is found that the larger bubble grows at the expense of the smaller one, indicating that the pressure is higher in the smaller bubble.

$(=2 \times 4\pi r^2)$. The work done against the pressure difference is $(p'' - p') \times A \times$ the distance through which the bubble shrinks:

$$\gamma \delta A = \gamma 8\pi 2r \delta r = (p'' - p') 4\pi r^2 \delta r$$

so that:

$$p'' - p' = 4\gamma/r \qquad \text{(for a bubble)}.$$

The corresponding formula for a drop of liquid (where there is only one surface to take into account, is the Young–Laplace equation:

$$p'' - p' = 2\gamma/r \qquad (5.6)$$

For a spherical surface, the important quantity is the radius of curvature, r, of the surface. If the surface is curved but not spherical it is necessary to specify the curvature a little more carefully. The curvature of any surface can be defined in terms of two radii, measured as indicated in Fig. 5.3. It turns out that, irrespective of where one chooses to draw the curve AXB, so long as CXD is drawn at right angles to it, the reciprocals of the two radii, r_1 and r_2, can be added to give a constant result, J, called the **mean curvature**. As the orientation of the curve AXB changes, the value of r_1 will, in general, also change, and at one particular orientation it will be a maximum, say R_1. The value of r_2 is then a minimum, R_2, and

$$\frac{1}{r_1} + \frac{1}{r_2} = J = \frac{1}{R_1} + \frac{1}{R_2}. \qquad (5.7)$$

R_1 and R_2 are called the principal radii of curvature of the surface; they can be positive or negative depending on whether the surface is concave or convex. Obviously, for a sphere, $r_1 = r_2 = r = R$ and $J = 2/r$.

In qualitative terms, the surface tension tends to compress a liquid droplet, increasing its internal pressure. The opposite situation arises with a concave liquid surface, as occurs in a capillary tube or with a bubble of air or vapour in a liquid. Here the pressure in the liquid is lower than the pressure inside the bubble. The Young–Laplace equation accounts for this situation since now R_1 and R_2 are negative, so that Δp is also negative. The radius is taken as positive if the corresponding centre of curvature lies in the phase in which p'' is measured, and negative in the converse case.

For a liquid in equilibrium, Δp must be constant over all parts of its free surface, otherwise liquid flow would occur.† Since γ is also constant, it follows that, *in the absence of external fields*, all liquid surfaces are surfaces of constant mean curvature J.

---

†This is true only in the absence of an external field. For macroscopic surfaces (like the meniscus of water in a beaker) gravity can exert enough force to influence the curvature at different parts of the interface.

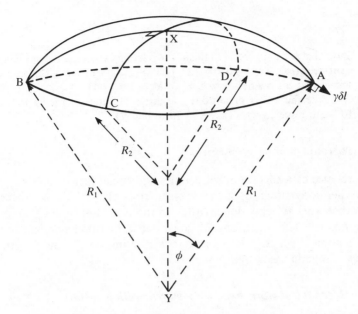

Fig. 5.3. Defining the meaning of the curvature of a surface at the point X. (See text.)

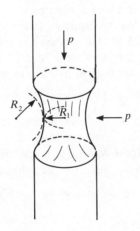

Fig. 5.4. A curved surface with no pressure drop across it because $1/R_1 = -(1/R_2)$. (After Adamson (1976).)

For an open film (for example a soap film on a wire frame) Δp is necessarily zero. This could mean $r_1 = r_2 = \infty$ (a flat film), but it could mean $r_1 = -r_2$ at all points on the surface, even though r_1 and r_2 change from point to point. This situation is illustrated in Fig. 5.4 which shows the shape expected for a soap film pulled between two open cylindrical pipes.

Exercises

5.2.1 Show that for a sphere of radius r, the relation between a volume change and a surface area change is $dA = (2/r)\,dV$. For a more general surface, it can be shown that the relation is: $dA = (1/R_1 + 1/R_2)\,dV$.

5.2.2 Calculate the excess pressure inside drops of water of radius 10^{-5} cm and 10^{-6} cm, respectively. (Take $\gamma = 70\,\text{mN m}^{-1}$.)

5.3 Thermodynamics of surfaces

The presence of a surface introduces an additional factor to be considered in the thermodynamics of such a system, since changes in the surface area imply that work is being done, either on the system or by the system on its surroundings. The equations of bulk phase thermodynamics thus need modification if surface changes are contributing significantly to the total energy change in the system.

5.3.1 Mechanical work done by a system with a surface

If a one-component, one-phase system expands by a volume dV against an external pressure p, the mechanical work done by the system is $dw = -p\,dV$ (that is, the work done is a negative contribution to the internal energy). If, however, the system contains two phases, we know from the Young–Laplace equation (5.6) that, unless the interface between the two phases has zero curvature, the pressure p'' and p' in the two phases will be different. Hence the total work done by the system will be

$$dw = -p'\,dV' - p''\,dV'' + \gamma\,dA. \tag{5.8}$$

Consider, for example, the work done during the evaporation of a liquid droplet of volume V'' having some arbitrary shape of constant mean curvature J (Fig. 5.5). If the total volume is V and the external pressure is p' then $V = V' + V''$. The work done on the system by its surroundings is

$$dw = -p'\,dV = -p'\,dV' - p'\,dV''$$

which may be written:

$$dw = -p'\,dV' - p''\,dV'' + (p'' - p')\,dV''. \tag{5.9}$$

Introducing the more general form of eqn (5.6):

$$dw = -p'\,dV' - p''\,dV'' + \gamma(1/R_1 + 1/R_2)\,dV''$$
$$= -p'\,dV' - p''\,dV'' + \gamma\,dA \tag{5.10}$$

using the result of Exercise 5.2.1. We can thus identify a work term for each of the phases and a 'surface work' term for the interface, without having to

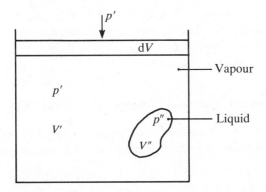

Fig. 5.5. Evaporation of a droplet whose surface has an arbitrary (but constant mean) curvature.

introduce a hypothetical surface piston or the stretching effect shown in Fig. 5.1.

5.3.2 Surface excess quantities

The presence of the surface affects virtually all of the thermodynamic parameters of a system. It is very convenient to think of a system containing a surface as being made up of three parts: two bulk phases, of volumes V' and V'', and the surface separating them. Any extensive thermodynamic property like the energy U can then be apportioned between these parts as follows. If the energies per unit volume in the two phases are u' and u'', then the total energy of the system due to contributions from the bulk phases must be $u'V' + u''V''$. The energy to be ascribed to the surface, U^σ, must then be given by

$$U^\sigma = U - u'V' - u''V''. \qquad (5.11)$$

Other surface quantities such as the surface entropy, S^σ, surface Helmholtz free energy, F^σ, and surface Gibbs free energy, G^σ, can be similarly defined.

It must be understood that these surface quantities can only be defined in terms of a particular model of the surface region. The model described above (and indicated by the superscript σ) is called the **Gibbs convention**. By dividing the total volume into the precise volumes V' and V'' where $V = V' + V''$ we have, in effect, constructed an imaginary system having the same properties as the real system but with two phases separated by an infinitesimally thin dividing surface and having constant densities in both phases up to that surface. This ideal system replaces the real system, with its finite dividing region where the density is rapidly, but not discontinuously changing. Other procedures are possible, especially for systems with flat interfaces, but they become difficult to manage with curved interfaces. In the Gibbs model we assume that the pressure in each of the regions is uniform, so if the surface is curved there will be a discontinuity in the pressure as we pass across the Gibbs surface.

5.3.3 Fundamental equations of surface thermodynamics

If a two-phase system, initially at equilibrium, is slightly disturbed, there will be infinitesimal changes induced in its properties and, in general,

$$U \to U + dU \qquad V' \to V' + dV'$$

$$S \to S + dS \qquad V'' \to V'' + dV''$$

$$n_i \to n_i + dn_i \qquad A \to A + dA \tag{5.12}$$

where n_i is the total number of moles of component i in the system. (This is called a **reversible process**: any **infinitesimal** change to a system in equilibrium must be reversible.) For the bulk phases we can write:

$$dU' = TdS' - p'dV' + \sum(\mu_i'dn_i') \tag{5.13}$$

and

$$dU'' = TdS'' - p''dV'' + \sum(\mu_i''dn_i'') \tag{5.14}$$

while for the Gibbs surface:

$$dU^\sigma = TdS^\sigma + \gamma dA + \sum(\mu_i^\sigma dn_i^\sigma). \tag{5.15}$$

Adding these three equations, and remembering that $U^\sigma = U - U' - U''$ and $S^\sigma = S - S' - S''$, we obtain a major governing equation of surface thermodynamics:

$$dU = TdS - p'dV' - p''dV'' + \gamma dA + \sum(\mu_i'dn_i' + \mu_i''dn_i'' + \mu_i^\sigma dn_i^\sigma). \tag{5.16}$$

The first term on the right is the heat absorbed by the system, the next three terms measure the mechanical work done on the system, and the last term sums over all components in the system and measures the chemical work done on the system.

Consider, for example, an infinitesimal process which occurs at a fixed temperature, T, in a two-phase system initially at equilibrium at constant total volume, V. Then the change, dF in the Helmholtz free energy of the system is given by

$$dF = d(U - TS) = dU - TdS$$

$$= -p'dV' - p''dV'' + \gamma dA + \sum(\mu_i' dn_i' + \mu_i'' dn_i'' + \mu_i^\sigma dn_i^\sigma). \tag{5.17}$$

But the system does no mechanical work at constant volume, so the sum of the first three terms must be zero. Also, since $dV' = -dV''$ we have:

$$-p''dV'' - p'dV' + \gamma dA = 0 = (p' - p'')dV'' + \gamma dA. \tag{5.18}$$

Then using the result quoted in Exercise 5.2.1 we can substitute for dA to obtain

$$(p'' - p')dV'' = \gamma(1/R_1 + 1/R_2)dV'' \tag{5.19}$$

from which the Young-La Place equation follows immediately. Note that we have derived it here purely on thermodynamic grounds.

Returning to eqn (5.17) we can introduce the condition $dn' + dn'' + dn^\sigma = 0$ to rewrite it in the form:

$$dF = \sum(\mu_i' - \mu_i^\sigma)dn_i' + \sum(\mu_i'' - \mu_i^\sigma)dn_i''. \tag{5.20}$$

Since F is a minimum in a closed isothermal system at fixed volume (that is, $dF = 0$) we have:

$$\mu_i' = \mu_i'' = \mu_i^\sigma(= \mu_i, \text{say}) \text{ for all components, } i. \tag{5.21}$$

Thus we may conclude that, at equilibrium, the chemical potential of every component in the system must be the same in both phases and at the interface.

5.4 Thermodynamic behaviour of small particles

The existence of a pressure difference across a curved interface (governed by the Young-Laplace equation (5.6)) has a number of important colloid chemical consequences. For very small particles (droplets or bubbles) the pressure difference may be so great that the chemical potential of the material is affected. Taking $\gamma = 70\,\text{mN m}^{-1}$ and a (spherical) drop radius of 50 nm, eqn (5.6) gives for the pressure difference ($2 \times 70 \times 10^{-3}/5 \times 10^{-8}$) Pa $\approx 28\,\text{atm}$. Such a pressure, applied to a liquid, will raise its chemical potential (and hence its vapour pressure) by a measurable amount. That can even alter the position of equilibrium of a chemical reaction.

The same excess pressure inside a gas bubble can be interpreted as a reduced pressure in the adjoining liquid and will cause a lowering of its vapour pressure. If the curvature is caused by the fact that the liquid-vapour interface is being formed as a meniscus in a small capillary, the same lowering of vapour pressure occurs, with important consequences for adsorption in porous solids. An idea of the magnitude of the vapour pressure change can be obtained from Kelvin's original derivation of a formula for the height to which a liquid will rise in a capillary.

5.4.1 Capillary rise and the Kelvin equation

When a capillary is dipped into a liquid, the liquid is drawn into the capillary by the forces of adhesion between the liquid and the capillary walls: we say that the liquid **wets** the capillary. The wetting proceeds until the force drawing the liquid up the tube is balanced by the downward force of gravity. Consider the system shown in Fig. 5.6 which is totally contained so that the space above the liquid contains only the vapour. For thermodynamic equilibrium to hold, the liquid and the vapour must be in equilibrium both at the flat liquid surface and the curved upper meniscus. Since the hydrostatic pressure (in the vapour phase) is lower at the upper meniscus by $\rho_v gh$ than it is at the plane liquid surface, it follows that the equilibrium vapour pressure at the meniscus is lower than at the plane liquid surface by this amount (ρ_v is the density of the vapour).

The Young–Laplace equation gives, at the meniscus:

$$\Delta p = (p^0 - \rho_v gh) - (p^0 - \rho_1 gh) = (\rho_1 - \rho_v)gh = \gamma(2/r) \quad (5.22)$$

so that

$$\rho_v gh = \gamma\left(\frac{2}{r}\right)\left(\frac{\rho_v}{\rho_1 - \rho_v}\right) \quad (5.23)$$

or

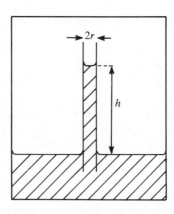

Fig. 5.6. Capillary rise of a wetting liquid in a cylindrical capillary tube contained in a closed, isothermal chamber. The space above the liquid contains only the vapour.

$$p'' - p^0 = (2\gamma/r)\left[\rho_v/(\rho_l - \rho_v)\right] \tag{5.24}$$

where p'' is the equilibrium vapour pressure at the meniscus and p^0 is the equilibrium vapour pressure above the flat liquid surface. Equation (5.24) is the original (approximate) form of the **Kelvin equation** which we will find, in its more general form, has a number of important applications. The way a wetting liquid is sucked into a capillary is itself of very great significance and we will discuss it in some detail in §5.7. It is involved in such diverse situations as the mopping up of liquid with a towel, the entry of water into soils, the way ink dries on paper, or paint covers a wooden surface, or trees manage to provide water to their leaves.

To derive a more general form of the Kelvin equation we begin with eqn (5.6) for a single pure substance, in the form of a sphere of radius r. Then $p'' - p' = 2\gamma/r$ and $\mu'' = \mu' = \mu$, where, by convention, the double prime and prime refer to the phase on the concave and convex side of the interface respectively. (Remember that $2/r$ can be replaced by $(1/R_1 + 1/R_2)$ for a general (nonspherical) interface.) The sign of r is positive for droplets (which therefore have an increased vapour pressure) and is negative for a concave meniscus.

For an infinitesimal process applied to a system initially at equilibrium, we have

$$dp'' - dp' = d(2\gamma/r) \tag{5.25}$$

and

$$d\mu'' = d\mu' = d\mu. \tag{5.26}$$

We now need to make use of a basic equation from thermodynamics (the Gibbs–Duhem equations) which takes the following form in each phase:

$$\begin{aligned} \bar{S}'dT - \bar{V}'dp' + d\mu' = 0 \\ \bar{S}''dT - \bar{V}''dp'' + d\mu'' = 0 \end{aligned} \tag{5.27}$$

when applied to a pure substance (where $\bar{S} = S/n_i$ and $\bar{V} = V/n_i$ are the molar entropy and volume respectively). The more general form of eqn (5.27) is derived from eqn (5.13) by the following argument.

Consider a system of several components in a single phase. If the size of the system is increased in the ratio $(1 + d\chi):1$, *whilst keeping T and P and the composition unchanged*, then for every component, $dn_i = n_i d\chi$ and the magnitude of each of the extensive variables is increased in the same proportion: $dU = Ud\chi$, $dS = Sd\chi$, etc. Substituting in eqn (5.13) we have:

$$dU = Ud\chi = TSd\chi - PVd\chi + \sum \mu_i n_i d\chi. \tag{5.28}$$

Integrating this expression from $\chi = 0$ to 1 (which corresponds to doubling the size of the system) gives:

$$U = TS - PV + \sum \mu_i n_i. \qquad (5.29)$$

We can now take the complete differential of eqn (5.29)

$$dU = TdS + SdT - PdV - VdP + \sum \mu_i dn_i + \sum n_i d\mu_i \qquad (5.30)$$

and comparing this with eqn (5.13) shows that†

$$SdT - VdP + \sum n_i d\mu_i = 0. \qquad (5.31)$$

This is the general form of the Gibbs–Duhem equation. What it says is that of the $m + 2$ variables $(\mu_1 \ldots \mu_m, P, \text{ and } T)$ only $m + 1$ are independent and the remaining one is fixed by eqn (5.31) (Atkins (1982, Ch. 8) gives some examples of the use of this equation for dealing with mixtures and solutions.) Equation (5.27) follows immediately from (5.31) for a one-component system. It expresses in mathematical terms the fact that in such a system, once the temperature and pressure have been fixed, the chemical potential is then fixed.

Equations (5.25)–(5.27) are the fundamental relations from which a large number of important results can be deduced. For example, at constant temperature we have (from (5.26) and (5.27)): $\bar{V}' dp' = \bar{V}'' dp''$ and so, from (5.25):

$$d(2\gamma/r) = ([\bar{V}' - \bar{V}'']/\bar{V}'')dp' \qquad (5.32)$$

$$= ([\bar{V}' - \bar{V}'']/\bar{V}')dp''. \qquad (5.33)$$

Equations (5.32) and (5.33) may be regarded as the most general forms of the Kelvin equation.

5.4.2 Applications of the Kelvin equation

(a) Drops of liquid in a vapour In this case (″) refers to the liquid and $\bar{V}'' \ll \bar{V}'$ ($\approx RT/p'$). Then from eqn (5.32):

$$d(2\gamma/r) = (RT/\bar{V}'')(dp'/p') - dp'. \qquad (5.34)$$

Integrating from $r = \infty$ (a flat surface where the equilibrium vapour pressure is p^0) to some finite radius where the equilibrium vapour pressure is p' we have

$$\ln(p'/p^0) = (\bar{V}''/RT)[2\gamma/r + (p' - p^0)] \text{ (droplets)} \qquad (5.35)$$

---

† This procedure is a bit puzzling at first encounter. It is simply a way of demonstrating a general property of the usual variables in a thermodynamic system and the constraints which are imposed on them by their mathematical definition.

or

$$\ln(p'/p^0) \approx 2\gamma \bar{V}''/rRT \qquad \text{if } \bar{V}'' \ll \bar{V}'.$$

Equation (5.35) is the exact form of Kelvin's equation. It shows that the equilibrium vapour pressure of the drop is higher than that of the flat liquid surface and the difference becomes larger as the drop becomes smaller. It follows, therefore, that if a number of droplets of uniform radius are initially in equilibrium with a surrounding vapour then that equilibrium must be **unstable**. Why? Because if one of the drops (as a result of some small fluctuation) began to grow slightly, its vapour pressure would fall and so it would continue to grow further. Conversely, if a droplet undergoes a slight evaporation, it will become smaller and its vapour pressure will increase, so that its evaporation will continue. In a mixture of drops of different radius, the large ones must always grow at the expense of the small ones just as the large bubble in Fig. 5.2 grows at the expense of the small bubble, and for similar reasons). Another way of putting the situation is to say that for a given vapour pressure there is a critical droplet size above which the drops will grow and below which they will evaporate. When a vapour is cooled so that it becomes supersaturated, condensation to the liquid cannot occur until there are formed some nuclei of the liquid which are large enough to continue to grow at the prevailing vapour pressure. That idea provides the basis of the theory of **homogeneous nucleation** which we will examine in §5.6.

An expression analogous to eqn (5.35) can be written for the solubility of a solid crystal in a surrounding liquid:

$$\ln \frac{C(r)}{C_\infty} = \frac{2\gamma V_m''}{rRT} \qquad (5.36)$$

where V_m'' is the molar volume of the solid, $C(r)$ and C_∞ are the solubilities of the particles and the bulk solid respectively, and γ is the interfacial surface energy per unit area. Although crystals tend to have flat faces it turns out that for small crystals it is the equivalent spherical radius which is required for eqn (5.36) (Hunter 1987). This equation again shows that larger crystals tend to grow at the expense of smaller ones, provided that the solid is sufficiently soluble to allow some transfer of solute from one to the other. One problem in applying it in any real system is our paucity of knowledge about the value of γ for the solid/liquid interface. In fact it is more useful as a means of estimating γ from the measured solubility than vice versa.

(b) Bubbles in a liquid In this case the phase indicated by the double prime is the gas and so we have (Exercise 5.4.2):

$$\ln p''/p^0 = (\bar{V}'/RT)\left[(p'' - p^0) - 2\gamma/r\right] \quad \text{(bubble or concave meniscus)}$$
$$(5.37)$$

$$\ln p''/p^0 \approx -(2\gamma\bar{V}'/rRT) \quad (\text{if } \bar{V}' \ll \bar{V}'').$$

Hence the vapour pressure inside the bubble is smaller than the value for a flat surface. This explains the phenomenon of superheating of liquids above the normal boiling point. Remember that there must be an excess pressure inside the bubble and that excess is larger the smaller the bubble is. But the vapour pressure of the liquid is smaller inside the bubble than it would be over a flat surface. Boiling can therefore only occur if either (i) the liquid can vaporize into a pre-existing bubble of reasonable size, or (ii) the temperature is raised sufficiently so that the equilibrium vapour pressure, even inside a small bubble, is large enough to allow it to continue to grow.

Case (i) is more usual. When water is heated, for example, tiny bubbles of air are released as the solubility of the air decreases with rise in temperature. Those bubbles form the nuclei into which water can vaporize at high enough temperature. If the water has been previously boiled, so that it no longer contains dissolved air, the boiling process becomes more difficult and the liquid may superheat and then boil very abruptly (causing 'bumping'). That is serious enough on the laboratory or domestic scale; it can be catastrophic on an industrial scale.

Case (ii) is called homogeneous nucleation (§5.6) and it can occur only at temperatures well above the normal boiling point. For water, for example, it requires a temperature of almost 200 °C to induce boiling at atmospheric pressure if rigorous efforts are made to exclude any extraneous nucleation bubbles and one relies entirely on the formation of bubbles of pure vapour.

Note that the changes in vapour pressure due to surface curvature become important only for very highly curved surfaces (Table 5.2).

5.4.3 *Effect of temperature on vapour pressure: the Thomson equation*

We now want to calculate the effect of temperature change on the vapour pressure inside a bubble. We must, therefore, derive the analogue of the Clausius–Clapeyron equation for the equilibrium between two phases across

Table 5.2 Influence of radius of curvature on the equilibrium vapour pressure above a spherical water surface, calculated from eqns (5.35) and (5.37) assuming that γ is independent of r

r (nm)	p'/p^0 (droplet)	p''/p^0 (bubble)
1000	1.001	0.999
100	1.011	0.989
10	1.115	0.897
1	2.968	0.337

a curved interface. This result was first derived by J. J. Thomson (the discoverer of the electron). We can simplify the problem by assuming that the latent heat of vaporization, ΔH_{vap}, per mole for the liquid is independent of curvature; what little change does occur only affects the behaviour at the very smallest sizes.

(a) Liquid drop suspended in its vapour From eqn (5.27), with double primes referring to the liquid, we have $dp' = 0$ (since the external pressure is constant) and so:

$$(\bar{S}' - \bar{S}'')dT + \bar{V}''dp'' = 0 \tag{5.38}$$

and

$$dp'' = d(2\gamma/r). \tag{5.39}$$

Setting $\bar{S}' - \bar{S}'' = \Delta H_{vap}/T$ and integrating from $r = \infty$ to some finite value r we obtain:

$$\ln T/T_0 = -(2\bar{V}''\gamma/\Delta H_{vap}r) \tag{5.40}$$

Table 5.3 Influence of curvature on the equilibrium temperature of droplets and bubbles. (From Defay *et al.* (1966).)

(a) Droplet of water in water vapour at 1 atm pressure

r(m)	T/T_0	T (K)	t(°C)
∞	1	373	100
10^{-5}	0.999995	372.998	99.998
10^{-6}	0.99995	372.98	99.98
10^{-7}	0.99948	372.807	99.81
10^{-8}	0.99485	371.08	98.08
5×10^{-9}	0.98906	368.92	95.92

(b) Bubble of water vapour in water at 1 atm pressure

r (m)	T/T^0	T (K)	t(°C)
∞	1	373	100
10^{-5}	1.0088	376	103.3
10^{-6}	1.0574	394.3	121.4
10^{-7}	1.2037	449	176
10^{-8}	1.461	545	272*

*At this temperature the surface tension has fallen from 55.5 to 18.4 mN m^{-1} and the latent heat from about 40 down to 31 kJ mol^{-1}.

which is the Thomson equation. We have assumed that both $\bar{V}''$ and ΔH_{vap} are unaffected by the temperature change, which is reasonable as the change is quite small in this case. Equation (5.40) allows us to estimate the temperature, T, at which small droplets of a liquid would boil, compared to the boiling point T_0 of the bulk liquid. Table 5.3 shows that as the drop size decreases the boiling point becomes lower, but the effect is quite small.

A more significant application of this equation occurs in particle physics. We can look on eqn (5.40) as describing the temperature at which condensation would occur on a small droplet, compared to the normal (bulk) condensation temperature. The equation shows that, as the droplets get smaller and smaller, so the temperature to which the vapour must be lowered to produce condensation also gets lower. This is the phenomenon of **supercooling**. Supercooled (that is **supersaturated**) water vapour is used in the **Wilson cloud chamber** to detect the passage of nuclear particles. The particles ionize the air in the chamber and this aids the formation of nuclei of sufficient size to permit further deposition (that is, drop growth) at the temperature of the chamber.

(b) Bubble immersed in a liquid   In this case double prime refers to the gas phase and $dp' = 0$ again. Assuming ideal behaviour for the gas ($p''\bar{V}'' = RT$) and using the Young–Laplace equation (5.6) we obtain from eqn (5.27) (Exercise 5.4.5):

$$\Delta H_{vap}\left(\frac{1}{T_0} - \frac{1}{T}\right) = R \ln\left(\frac{2\gamma/r + p'}{p'}\right). \tag{5.41}$$

This form of Thomson's equation gives the temperature at which a bubble of vapour of radius r can exist in equilibrium inside a liquid. For small values of r it follows that $T > T_0$ and, indeed, the temperature can be much higher than the normal boiling point as we noted above. Recall that there has to be an excess pressure inside the bubble so eqn (5.41) calculates how high the temperature must be raised to achieve that higher vapour pressure.

Table 5.3 gives values for the equilibrium temperatures for drops and bubbles of various sizes. Notice that the effect on the equilibrium of bubbles is much larger than on that of drops. That is not surprising when one considers that the effect is due to the excess pressure inside the sphere; the effect of pressure on the chemical potential of a gas is much greater than its effect on a liquid. (Note that the figures in the table for bubbles are derived using a rather more elaborate equation, allowing for some variation of ΔH_{vap} with temperature, because the temperature range is quite wide.)

5.4.4   Application of the Kelvin and Thomson equations to solid particles

Small solid particles can be treated as spheres with some equivalent radius and the resulting excess pressure experienced by a solid will increase its chemical potential. The practical effects of that change can be very important. For example:

1. The vapour pressure of small crystals is greater than that of large crystals. In the presence of the vapour, large crystals will grow at the expense of small crystals.
2. Small crystals will melt at a temperature lower than the normal melting point. The melting point, T, of a small crystal will be given by

$$\ln(T/T_0) = -(2\gamma_{sl}\bar{V}_s)/(r\Delta H_{fus}) \qquad (5.42)$$

   where T_0 is the normal melting point at the same external pressure, γ_{sl} is the interfacial energy, $\bar{V}_s$ is the molar volume of the solid, and ΔH_{fus} is the molar heat of fusion.
3. The melting point of a substance solidified in the pores of an inert material will depend upon the size of the pores.
4. Small crystals may have a heat of fusion and a heat of sublimation smaller than the value for the bulk solid.

Point 2 is very important in the field of ceramics, where finely powdered materials are heated to high temperatures. The particles will usually be rough so that the crystals will have sharp protrusions (asperities) which have a small radius of curvature. At these points, **sintering** occurs. That is, a fusion and partial interdiffusion occurs at these localized centres at a temperature below the normal melting point. That binds the whole mass together into the final product. A similar process is a major cause of the friction between metal parts moving against one another: small rough spots on one surface can melt, and alloy with the other metal surface, solidify as the pressure is reduced, and then have to be torn from the surface as the metal parts continue to move with respect to one another. That is why so much energy is lost in the process.

Point 3 is important in the treatment of oil deposits: treating oil sands and shales to recover heavy crude oil depends on a knowledge of the capillary pressure to which the oil is subjected since this will determine its melting point. The problem is obviously more important for the extensive oil deposits which occur near the Arctic Circle.

Another important consequence of the excess pressure inside a small particle is its effect on a chemical equilibrium. Figure 5.7 shows the effect of adding fine particles of Ni powder to an equilibrium mixture of Ni foil and

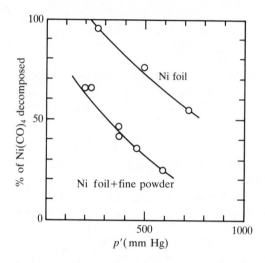

Fig. 5.7. Effect of particle size on the Ni–Ni(CO)$_4$ equilibrium (Adamson 1976).

Ni(CO)$_4$. The effect is quite dramatic and is a direct consequence of the higher chemical potential of the nickel when it is in the form of small particles.

Exercises

5.4.1 Use eqn (5.24) to estimate the fractional lowering of the vapour pressure (p''/p^0) above the meniscus of water in a capillary of radius 0.1 mm. (Note that r is negative in this case and assume that the vapour behaves ideally.) Assume that the liquid has a density of 1.00000 g cm^{-3}. To what height will the liquid rise. (Take $\gamma = 70$ mN m^{-1}.)

5.4.2 Establish eqn (5.37). Establish the connection between this form of the Kelvin equation and equation (5.24).

5.4.3 Discuss the function and mode of operation of 'boiling chips' (small pieces of porous ceramic material) as a means of avoiding 'bumping' in a boiling liquid. Why do they become useless for this purpose when they are left in the liquid as it cools and the liquid is reboiled?

5.4.4 Verify the entries in Table 5.2.

5.4.5 Establish eqn (5.41).

5.5 Behaviour of liquids in capillaries

The difference in pressure that occurs across a curved interface has many important consequences. If the radius of curvature of the liquid–gas interface is negative (as in a gas bubble in a liquid) then the pressure is lower in

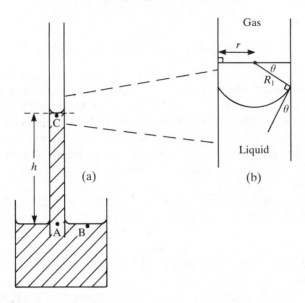

Fig. 5.8. Rise of liquid in a capillary tube of radius r. The pressure must be the same at points A and B.

the liquid than it is in the gas. We have, so far, interpreted this as an excess pressure inside the gas bubble. When the liquid is contained in a capillary tube (Fig. 5.8) it is more useful to think of the liquid pressure as being *negative* with respect to the gas pressure. A negative pressure (or suction pressure) of this kind is like a tension and the magnitude of that tension can be very high; it may be limited only by the tensile strength of the liquid.

5.5.1 Capillary pressure

Consider the liquid in Fig. 5.8. The height to which it will rise in the capillary is determined by the curvature of the meniscus. If the radius of the tube is sufficiently small, gravitational distortion can be neglected and we can estimate the radii of curvature, R_1 and R_2, of the interface as $R_1 = R_2 = r/\cos\theta$ (Fig. 5.8(b)) where θ is called the **contact angle**. (This is the angle between the liquid surface and the solid, measured in the liquid. We will discuss it in more detail in the next section. For the moment we assume that it can be observed and measured macroscopically with sufficient accuracy.) The pressure must be the same at points A and B and is *lower* at point C by an amount $\rho_l gh$. The pressure difference across the upper interface is then given by (compare eqn (5.22):

$$\Delta p = \rho_l gh - \rho_v gh = \gamma(1/R_1 + 1/R_2)$$

so that

$$h = (2\gamma\cos\theta)/(r[\rho_l - \rho_v]g) \approx (2\gamma\cos\theta)/r\rho_l g) \qquad (5.43)$$

where ρ_l and ρ_v are the densities of liquid and vapour respectively. This is the simple equation for the rise of a liquid in a narrow tube and it is frequently used for the measurement of surface tension, especially of pure liquids. It is really only practicable when $\theta = 0$ (that is when the liquid completely wets the solid). If the contact angle is non-zero it may be difficult to measure with sufficient accuracy and may be variable. We are concerned here with the practical consequences of eqn (5.43) rather than its application to the measurement of γ.

In normal laboratory measurements, the capillary radius is of the order of 1 mm and the rise, h, for water would be about $(2 \times 70 \times 10^{-3} \times 1)/(10^{-3} \times 10^3 \times 9.8)$ m ≈ 14 mm. For very tiny capillaries the rise can be very high indeed, and eqn (5.43) has been used to explain the transport of water to the tops of very tall trees. If capillary pressure alone were to account for this movement for a tree of, say, 100 m height, the radius would have to be about 1.5×10^{-7} m or about 150 nm. That is by no means impossible as there are apertures of such dimensions in the leaf surface. It is not necessary for the tube to be of the same radius throughout. Only the radius at the upper surface is important and the tube bore can be much wider lower down (Fig. 5.9). It should be clear, however, that in cases (b) and (c) the liquid will not rise of its own accord to height h because it will not be able to traverse the wider parts of the tube; the pressure drop across the interface disappears as the liquid surface moves through those regions. The situation in (b) and (c) is stable, however, if the liquid is sucked up to that height and then the suction pressure is removed.

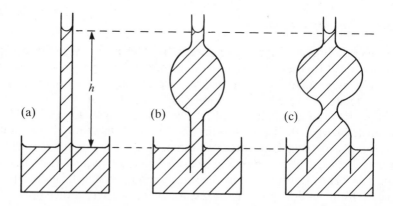

Fig. 5.9. Capillary equilibrium is affected only by the curvature at the interface and not by shape of the capillary at other levels (but see text).

The movement of liquid into narrow capillaries, against the action of gravity, has other important consequences. The capacity of a soil to retain water at the root zone of plants is determined by capillary (pore) structure and this depends chiefly on the particle size of the soil constituents. Very fine (clay) particles produce very small pores that can prevent water from draining away to the level of the underground water table. Too much clay, however, (that is, too heavy a soil) can make it difficult for the plant to extract the water since it must provide a still greater suction pressure to draw the water out.

These negative (suction) pressures are not limited to 1 atm. The calculation involving the tree corresponds to a negative pressure of about 10 atm. We can set an upper limit by the consideration that r cannot be less than molecular dimensions (say 5 nm) and taking $\gamma \approx 100\,\mathrm{mJ\,m^{-2}}$ as typical we have $\Delta p_{max} \approx 4 \times 10^7\,\mathrm{N\,m^{-2}} \approx 400\,\mathrm{atm}$—more than enough to have a profound effect on the properties of the material in the pores.

One final point about eqn (5.43) concerns the contact angle θ. We have discussed, for the most part, liquids for which $\theta = 0$, that is perfectly wetting liquids. Whether a liquid wets (spreads freely over) another liquid or a solid surface depends upon how strong the interaction is between the two materials, compared to the cohesive forces which hold the liquid together. (§5.7). Here we need only note that for a nonwetting liquid, $\theta > \pi/2$ and eqn (5.43) predicts, a *negative* value for h. In that case the meniscus is convex (for example, as in a liquid drop), the capillary pressure inside the liquid is *positive* and the surface of the liquid inside the capillary is *lower* than the free surface outside. That is the behaviour observed with mercury in a glass capillary, for which $\theta \approx 135°$. In order to make mercury enter a porous solid it is necessary to apply an *increased* positive pressure to the liquid (since in most cases it is nonwetting). By determining the volume of mercury taken up by a solid as a function of the increase in pressure, one can determine the pore size distribution, a procedure known as **mercury injection porosimetry**. Successive increases in pressure allow successively smaller pores to be entered (in accordance with the Young–Laplace equation (5.6). One must still assume a contact angle in order to relate R_1 and R_2 in that equation to the capillary radius, but the method is a very useful one for comparing different samples of the same or similar materials, where θ can be assumed not to vary greatly. It is an important technique for characterizing different catalyst support materials, where pore size has an important effect: in some cases the nature of the product is determined by how big a molecule can get access to the catalyst adsorbed on the surface of the pores.

We referred above to the effect of capillary pressure on the melting point of solid materials confined inside the pores of another solid. It should now be clear that one cannot make sensible predictions about the sign let alone the magnitude of such effects from a knowledge of the pore size alone. The effect of the contact angle is crucial for a liquid, since it determines whether the capillary pressure is positive or negative. But how can this apply to solids?

The problem with solids concerns not only the contact angle but the concept of surface energy itself. Most of the discussion to date has involved the assumption that the system has either reached equilibrium or that at least the surface energies involved are well defined. For the liquid–vapour interface that is usually a reasonable assumption. There may be some time effects connected with the diffusion of an adsorbate to the surface but these are minor compared to the problem with solids. As we noted in §5.2.1, a solid surface (especially one produced by fracture) may remain in a nonequilibrium state indefinitely, with an arbitrary surface energy, since it has no means of relaxation. We should not be surprised then to find that problems involving the surface tension or energy of solids give rise to considerable controversy. The solid–liquid–vapour interface is, however, so important in so many areas of colloid science that we must be prepared to make some attack on it, even if that involves some sacrifice of rigour. The insights so gained, however imperfect, can help us to order our view of the physical world and make reasonable projections from the known to the unknown, which is what much of the great game of science is all about.

5.6 Homogeneous nucleation

The Thomson equation allows us to calculate the temperature at which a drop or a bubble of a given size can be in equilibrium with the surrounding phase (assuming that the latter is very large). As we have already noted, that equilibrium is *unstable*; if the drop begins to grow by the random accumulation of a few more molecules†, its radius increases, its vapour pressure decreases and growth can continue with a further drop in free energy. That same unstable equilibrium is involved in the Kelvin equation: the radius of the drop that is in equilibrium with the surrounding vapour pressure is a critical radius, r_c, because smaller drops have higher vapour pressure and will spontaneously evaporate. If the critical drop accumulates a few more molecules it can go on increasing in size because its equilibrium vapour pressure is lower than the prevailing value. It is important to note that not all such droplets *will* increase in size. A significant fluctuation in the opposite direction can reduce r below r_c before the droplet has time to grow, and then the driving force is in the opposite direction. We can only say that, if there are a reasonably large number of such drops, then there is a high probability that some of them will stay on the high side of r_c long enough to

† This is called a **fluctuation** and such random (positive and negative) changes are an important aspect of the kinetic molecular theory.

become so much larger than r_c that they will never evaporate.

The unstable nature of the equilibrium becomes clear if we write the equation for the free energy change involved in forming a small drop of liquid from the vapour:

$$nA\,(\text{gas},p) \leftrightarrow A_n(\text{small liquid drop}).$$

We need to know the free energy before and after the formation of a drop of liquid from the (supersaturated) vapour in a one-component system at constant temperature and constant total overall volume, V. Before formation of the drop of liquid we can obtain the Helmholtz free energy by integrating the general equation for $\mathrm{d}F$:

$$\mathrm{d}F = \mathrm{d}(U - TS) = -p\mathrm{d}V - S\mathrm{d}T + \sum \mu_j \mathrm{d}n_j.$$

The result, for a single component at constant temperature, is

$$F_0 = n_i \mu_g(p_i') - p_i' V \tag{5.44}$$

where n_i is the initial number of moles present and p_i' is the initial pressure in the gas phase. After formation of the drop or drops, the free energy of the entire system is obtained by integrating eqn (5.17) at constant temperature to give

$$F = -p'V' - p''V'' + \gamma A + n'\mu_g(p') + n''\mu_l(p'') \tag{5.45}$$

where $V = V' + V''$, $n_i = n' + n''$, and μ_l is the chemical potential of the liquid.

The free energy of formation of the droplet phase is then

$$(\Delta F)_{T,V} = F - F_0. \tag{5.46}$$

Note that this is true irrespective of whether the process is an equilibrium or a nonequilibrium one, since it only involves the initial and final state functions.

If the volume of the gas phase is sufficiently large, the amount of material needed to form the droplet phase does not appreciably alter the pressure so

$$\mu_g(p_i') = \mu_g(p') \quad \text{and} \quad p' = p_i'. \tag{5.47}$$

Then from eqns (5.44)–(5.47):

$$(\Delta F)_{T,V} = n''\,(\mu_l(p'') - \mu_g(p')) - V''(p'' - p') + \gamma A. \tag{5.48}$$

This equation can be used to calculate ΔF for the formation of a droplet of any size, from our knowledge of the influence of drop size on $p'' - p'$ (the Young–Laplace equation (5.6)) and on $\mu_l - \mu_g$ (the Kelvin equation (5.35)).

For example, the Young–Laplace equation gives

$$V'' (p'' - p') = (4/3)\pi r^3 \times (2\gamma/r) = (2/3)(4\pi r^2 \gamma) = 2\gamma A/3. \tag{5.49}$$

Furthermore, at the critical radius, the liquid drop and the vapour are in equilibrium so that $\mu_l(p'') = \mu_g(p')$ and hence, from eqn (5.48):

$$(\Delta F)_{\text{crit}} = -V''(p'' - p') + \gamma A = (1/3)\gamma A = 4\pi r_c^2 \gamma/3 \tag{5.50}$$

where r_c is the critical radius.

It can be shown (Exercise 5.6.1) that for larger or smaller droplets

$$(\Delta F)_{T,V} = -n'' RT\ln(p'/p^0) - V''(p^0 - p') + \gamma A \tag{5.51}$$

where p^0 is the equilibrium vapour pressure of the liquid over a plane interface at the temperature of the system. Provided the degree of supersaturation (p'/p^0) is not too large we may neglect the second term on the right and write

$$(\Delta F)_{T,V} = -n'' RT\ln(p'/p^0) + \gamma A. \tag{5.52}$$

The first term is negative under supersaturation conditions and when it becomes large enough it dominates over the second term, so that ΔF can become negative. To express this in terms of the drop radius we note that $n'' = (4/3)\pi r^3/\bar{V}''$ and so:

$$(\Delta F)_{T,V} = -\frac{4}{3}\frac{\pi r^3}{\bar{V}''} RT\ln\frac{p'}{p^0} + 4\pi r^2 \gamma. \tag{5.53}$$

Then

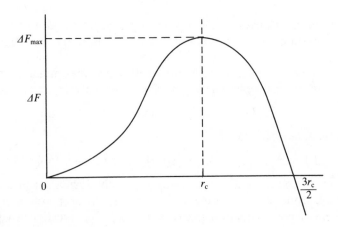

Fig. 5.10. Variation with radius of the free energy of formation of a nucleus in a supersaturated homogeneous phase.

$$\left(\frac{\partial \Delta F}{\partial r}\right)_{T,V} = -\frac{4\pi r^2}{\bar{V}''} RT \ln \frac{p'}{p^0} + 8\pi r\gamma$$

$$= 4\pi r^2 \left(\frac{2\gamma}{r} - \frac{RT}{\bar{V}''} \ln \frac{p'}{p^0}\right)$$

(5.54)

which is obviously zero when Kelvin's equation (5.35) is satisfied. Note that this is a maximum value for ΔF corresponding to an unstable equilibrium and that these positive values of ΔF correspond to nonspontaneous processes. The dependence of ΔF on r is plotted schematically in Fig. 5.10. The initial rise is a quadratic (r^2) function where γA dominates until r becomes large enough for the first term in eqn (5.53) to take over.

The value of ΔF_{max} is given by (Exercise 5.6.2)

$$\Delta F_{max} = \Delta F_{r=r_c} = \frac{16\pi\gamma^3 M^2}{3\rho_1^2 [RT \ln(p'/p^0)]^2}$$

(5.55)

where M is the molar mass and ρ_1 the density of the liquid. ΔF can be regarded as an activation energy for formation of a nucleus of sufficient size. The rate of nucleation, Ω, is therefore given by an Arrhenius type equation:

$$\Omega = Z \exp(-\Delta F_{max}/kT)$$

(5.56)

where Z is determined by the collision frequency of the gas molecules and so will increase with $(p')^2$. A detailed calculation of the terms in this equation (Adamson 1976) shows that, as the degree of supersaturation (p'/p^0) gradually increases, a point is reached (at about $p'/p^0 \approx 4.2$ for water at $0\,°C$) where the value of Ω suddenly increases extremely rapidly. This turns out to be very close to the experimentally observed degree of supersaturation which induces formation of a fog in water vapour free of dust particles or other nucleation sites.

The same effect can be achieved by lowering the temperature, which also increases the degree of supersaturation. Over the early stages of temperature reduction, the rate of condensation increases relatively slowly with decrease of temperature, but eventually a temperature is reached where the chance of forming a nucleus of critical radius, (a **critical** nucleus) by spontaneous fluctuations in molecular association, increases very rapidly with a very small further reduction in temperature. The temperature at which this rapid change occurs is called the **homogeneous nucleation temperature**. To investigate such effects it is necessary to rigorously exclude any dust particles which act as sites onto which nucleation could begin more easily than by the formation of a critical nucleus.

Exercises

5.6.1 Establish eqn (5.51) using the fact that $\mu_1(p'') = \mu_1^0 + \int \bar{V}'' \, dp$ at constant temperature. The limits for the integral are from p^0 to p''; μ_1^0 is the standard chemical potential of the liquid at pressure p^0 which can be taken as the equilibrium vapour pressure over a flat interface at the temperature in question. $\bar{V}'' = \bar{V}_1$ is the molar volume of the liquid.

5.6.2 Verify eqn (5.55).

5.6.3 Show that there is no energy barrier to growth for $r > 3r_c/2$, as indicated in Fig. 5.10.

5.6.4 Estimate the size of a critical nucleus from the fact that the critical supersaturation is 4.2 for water at 0 °C. (Take $\gamma = 70 \text{ mN m}^{-1}$.) What is the volume of

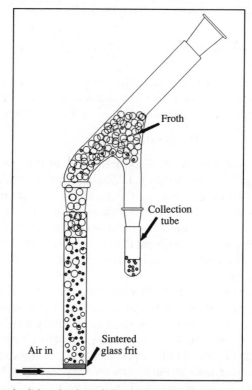

Fig. 5.11. The principle of mineral flotation as applied in the laboratory, in the Hallimond tube. A stream of air bubbles issues from a porous glass frit, into a tube containing the mineral particles and gangue (worthless) material. An additive is dissolved in the solution and it adsorbs on the mineral grains to make them nonwetting (that is, hydrophobic). This allows the air bubbles to become attached to the mineral grains as they rise. The mineral grains can be collected from the froth at the top of the cell. The gangue (black circles) is not carried up in the froth because the air bubbles do not become attached to those grains. (See §10.3 for further details.)

the critical drop? How many molecules will it contain? (You should get about 70 molecules. Rather surprisingly, the macroscopic concept of surface tension still works quite well even for such tiny droplets. Physicists even used a comparable notion for discussing the energy of a single atomic nucleus.)

5.7 Contact angle and wetting behaviour

When a drop of liquid is placed on a solid surface, the contact angle is a measure of the competing tendencies of the drop to spread out and cover the solid surface and to round up so as to minimize its surface area. Rounding up may seem to be the natural process, since it also reduces the area of liquid/solid contact but that neglects the fact that the solid/vapour interface is thereby increased.

The contact angle measures the **wetting** tendency of a liquid on a solid or another liquid, and is of particular importance in processes like fabric treatment (conditioning, dyeing, and cleaning), insecticide application, and mineral flotation. An insecticide like a cattle dip, for example, will be much less effective if is not able to penetrate the coat to get down to the skin level and that means making it wet the hair to the right extent; if it wets too well it will also drain too well. Separating mineral grains, on the other hand, requires that they be made nonwetting so that an air bubble will stick to them and float them to the surface where they can be carried away in the foam for further treatment (Fig. 5.11).

5.7.1 *Adhesion, cohesion, and wetting*

The **cohesion** in a liquid measures how hard it is to pull apart. The **work of cohesion**, W_{AA}, is the (reversible) work, per unit area, required to break a column of liquid into two parts, creating two new **equilibrium** surfaces, and separating them to such a distance that they are no longer interacting with one another (a few micrometres is sufficient). From Fig. 5.12:

$$W_{AA} = 2\gamma_{AV} \tag{5.57}$$

where γ_{AV} is the surface tension of liquid A in contact with its vapour.

By the same token, the work of adhesion per unit area between two different immiscible liquids A and B is given by the Dupré equation:

$$W_{AB} = \gamma_{AV} + \gamma_{BV} - \gamma_{AB} \tag{5.58}$$

since two new liquid–vapour interfaces are created and the original interface disappears. In these equations γ_{AV} represents the quantity $(\partial F/\partial A)_{T,V,n}$ (easily derived from eqn (5.17)). In what follows we will be concerned only with planar interfaces and for such systems many workers prefer to introduce the more familiar (to chemists) Gibbs free energy function. This is permissible since the pressure can now be kept constant throughout the

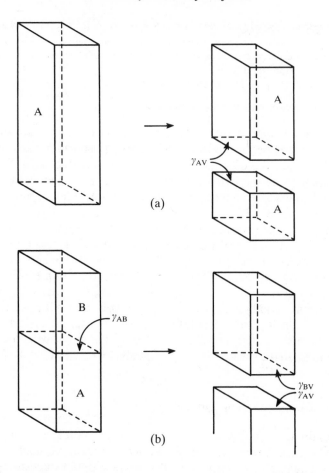

Fig. 5.12. Illustrating (a) the work of cohesion, W_{AA} in a liquid and (b) the work of adhesion W_{AB} between two different liquids. Breaking the column creates two new interfaces in the first case. In the second case it also destroys one.

entire system, because there are no curved surfaces. The alternative definition of γ is then

$$\gamma_{AV} = (\partial G_A / \partial A)_{p, T, n}. \tag{5.59}$$

When an oil drop is placed on a clean water surface it may form a lens as shown in Fig. 5.13. The overall shape is described by the Young–La Place equation (5.6), subject to the condition that there must be a balance of forces along the line of contact between the two liquids and the vapour above them:

$$\gamma_{wv} \cos \theta_3 = \gamma_{ov} \cos \theta_1 + \gamma_{wo} \cos \theta_2. \tag{5.60}$$

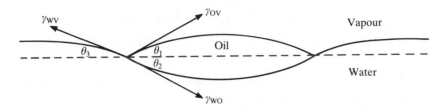

Fig. 5.13. Initial shape of an oil lens on a water surface. The shape may change with time if the oil is able to spread and/or if the two liquids are to some extent mutually soluble.

A more interesting version of this equation applies when the lower liquid is very dense compared to the upper liquid (like water on mercury). In that case both θ_2 and θ_3 approach zero and we have:

$$\gamma_{MV} \approx \gamma_{MW} + \gamma_{WV} \cos \theta_1 \qquad (5.61)$$

where M refers to mercury, W to water, and V to the equilibrium vapour. This corresponds to the situation one would expect if the liquid were sitting on a solid (nondeformable) substrate. Equation (5.61) then takes the form

$$\gamma_{SV} = \gamma_{SL} + \gamma_{LV} \cos \theta \qquad (5.62)$$

where S, V, and L refer to the solid, vapour, and liquid respectively. Equation (5.62) was put forward by Thomas Young in 1805 and goes by his name. Its simplicity belies the very considerable argument it has generated over the last couple of hundred year (Hunter 1987). For the moment we will accept it as true subject to the proviso that the γ values are **equilibrium values** of surface free energy, obtained after the adsorption processes have ceased. In particular, at the solid/vapour interface, the surface free energy will not be that of the pure solid, γ_s^0 but will be some lower value due to the adsorption of some of the vapour. (The adsorption process occurs *because* it lowers the free energy of the surface; we will have more to say about this in the next chapter.) The lowering of surface free energy due to adsorption of a film of adsorbate is

$$\Pi_e = \gamma_s^0 - \gamma_{SV}. \qquad (5.63)$$

It has the dimensions of an energy per unit area or a force per unit length but it is often interpreted as a two-dimensional pressure called the **spreading pressure** of the adsorbate. The reason for that sort of interpretation will become clearer when we discuss the behaviour of adsorbed films in the next chapter.

In terms of the surface energy γ_s^0 of the pure solid (in contact with its own vapour but not with liquid L) we have:

$$\gamma_{LV} \cos\theta = \gamma_s^0 - \Pi_e - \gamma_{SL}. \tag{5.64}$$

The work of adhesion, W_{SL}, between a solid and a liquid can then be defined, by analogy with equation (5.58), as:

$$W_{SL} = \gamma_s^0 + \gamma_{LV} - \gamma_{SL} \tag{5.65}$$

which, using eqn (5.64), becomes:

$$W_{SL} = \gamma_{LV}(1 + \cos\theta) + \Pi_e. \tag{5.66}$$

An important question relating to Fig. 5.13 is whether the liquid lens will spread out over the water surface or remain in the form of a lens; in other words will the oil wet the water? In the limiting case where θ_1, θ_2, and θ_3 approach zero, the lens becomes a film, and spreading is determined by the quantity

$$dG = \left(\frac{\partial G}{\partial A_{wv}}\right)dA_{wv} + \left(\frac{\partial G}{\partial A_{ov}}\right)dA_{ov} + \left(\frac{\partial G}{\partial A_{wo}}\right)dA_{wo}. \tag{5.67}$$

If some spreading occurs, the spreading film will increase its contact with both the water and its own vapour, but it will cover some of the WV interface so that:

$$dA_{wo} = dA_{ov} = -dA_{wv}. \tag{5.68}$$

Also we know that $(\partial G/\partial A_{wv})_{P,T} = \gamma_{wv}$ with similar expressions for the other terms in eqn (5.67). For spreading to occur spontaneously it is necessary for dG to be negative at constant temperature and pressure when dA_{wo} is positive. This will be so when the coefficient $S_c = -(dG/dA_{wo})$ is positive. S_c is called the **spreading coefficient** and from eqns (5.67) and (5.68):

$$S_c = \gamma_{wv} - \gamma_{ov} - \gamma_{wo}. \tag{5.69}$$

From the definitions of work of adhesion and cohesion given in eqns (5.57) and (5.58) it is obvious that

$$S_c = W_{ow} - W_{oo}. \tag{5.70}$$

Thus the spreading coefficient measures the difference between the adhesion of oil to water and the cohesion of the oil to itself. It is in this sense that we use the term **wetting**. A liquid will wet another material (liquid or solid) if its own work of cohesion is less than the work of adhesion between it and the substrate. A curious example is that of benzene on water. In that case S_c is positive for benzene on clean water but negative for benzene in contact with a saturated solution of benzene in water. Thus, benzene will initially spread on water but will then round up into lenses as the water becomes saturated with benzene.

Water wets many surfaces very well and this has some important consequences. Sometimes we want to prevent wetting (as in 'waterproofing' a
fabric or floating a mineral particle with an air bubble). In other cases, wetting must be promoted. For example, garden soil sometimes becomes
infected with fungus which makes the soil particles so hydrophobic that
water cannot enter the soil. To overcome this and similar problems it is
necessary to modify the surface energies γ_{SL}, γ_{SV}, and γ_{LV} and this is done
by adsorbing suitable materials onto one or other of the surfaces. Substances
that adsorb readily to surfaces are called **surfactants**. We have mentioned
them before and will discuss their action a little further in the next chapter.

References

Adamson, A. W. (1976). *Physical chemistry of surfaces*, (3rd edn). Wiley, New York.

Atkins, P. W. (1982). *Physical chemistry*, (2nd edn), Ch. 8. Oxford University Press.

Aveyard, R. and Haydon, D. A. (1973). *Introduction to the principles of surface chemistry*, p. 70. Cambridge University Press.

Defay, R., Prigogine, I., Bellemans, A. and Everett, D. H. (1966). *Surface tension and adsorption*, p. 242. Longmans Green, London.

Hunter, R. J. (1987). *Foundations of colloid science*, Vol. I, pp. 269–72. Oxford University Press.

ADSORPTION AT INTERFACES

6.1 Adsorption at the solid–gas interface

The earliest electric light bulbs, such as those introduced by Thomas Edison, used as a light source a hot carbon filament inside an evacuated glass bulb. These were subsequently superseded by the metal tungsten as filament material and the bulbs were filled with a gas (at first nitrogen but later mostly argon with a little nitrogen), to reduce the amount of metal evaporation at the high temperatures involved. To produce an efficient bulb it was necessary to understand the details of what was happening at the metal surface in contact with the gas. The most prolific and effective researcher in this area, in the first decades of this century, was Irving Langmuir, who made an extensive study of the adsorption of gases onto metal surfaces and how the amount adsorbed was affected by the nature of the metal and the gas pressure, composition, and temperature.

From that early work a huge area of study has developed, spurred on by the scientific and technological problems involved in the electronics industry, first through the development of the electronic valve or vacuum tube and subsequently the transistor. There is now a wide variety of tools available with which to examine the details of the surface structure and the way it is changed by reaction with an adsorbing molecule. To do justice to that area would, however, take us too far afield. Some discussion of gas adsorption on solids is given in the standard physical chemistry texts (Atkins 1982); we will concentrate here on a few aspects of relevance to colloid chemistry.

One important application of the study of adsorption is to the determination of the total surface area of a finely divided solid. A catalyst, for example, is often produced in the form of a fine powder with the active material on its surface. Its effectiveness is then determined largely by how much area there is to adsorb the reactants and generate the products. The area of a solid can be measured if we can determine the number of gas molecules of known size required to form a single layer on the surface.

Before it became possible to study the microscopic details of the adsorption process and the molecular interactions which were going on at a metal surface, the main insights were provided by examining the macroscopic behaviour, in particular how fast molecules of gas became adsorbed and how much was adsorbed as a function of gas pressure and temperature. Note that we use the term **adsorption** to indicate that the molecule is stuck to the surface, as distinct from **absorption** where molecules would go into the interior of a substance.

It is also useful to distinguish two types of adsorption: **physical** and **chemical**, referred to as **physisorption** and **chemisorption** respectively. In the first case, the gas molecule is bound to the solid surface by physical forces, chiefly the van der Waals force. In the second case a chemical bond is formed between the gas (called the **adsorbate**) and the underlying solid (called the **adsorbent** or **substrate**). A physisorbed gas molecule may have its bond structure slightly distorted but it retains its identity. A chemisorbed molecule normally breaks down to some extent and the fragments then interact with the surface atoms of the solid. The energy involved in the latter case (usually of the order of several hundred kilojoules per mole) is significantly higher than for physisorption (where the energy is more like $30–60 \, kJ \, mol^{-1}$). Physical adsorption increases as the temperature is lowered, whereas chemical adsorption normally decreases at low temperature because the activation energy required to break chemical bonds is no longer available.

Gases which are strongly bonded and relatively inert, like nitrogen, are more likely to be physisorbed on a metal surface. Oxygen, on the other hand, is more usually chemisorbed to form an oxide layer on a metal surface. Sometimes a gas has a special relation with a particular metal, as is for instance the case with hydrogen gas on palladium. There the interaction is so strong that the gas dissociates to individual atoms which then penetrate into the metal crystal.

6.1.1 The mechanics of adsorption

When a clean metal surface is brought into contact with a gas at a reasonable pressure, say of the order of 0.01 atm, the gas molecules will bombard the surface in large numbers. The number of collisions per unit area, at normal temperatures, can be calculated from the kinetic theory (Atkins 1982) and is about 3×10^{23} cm^{-2} s^{-1} at this pressure. Since each square centimetre of surface contains about 10^{15} atoms, each of those atoms will be struck by a gas molecule about 10^8 times per second. Some of those molecules simply bounce back off the surface, but at any particular pressure a certain fraction will remain. The extent of adsorption is measured by the **coverage** θ, which is given by

$$\theta = \frac{\text{Number of surface sites occupied}}{\text{Total number of surface sites}}. \qquad (6.1)$$

The rate at which the first adsorbed layer is built up decreases as the layer nears completion. There is also a tendency for the most highly active sites on the surface to be filled up first, and this too makes the rate fall off as the surface fills up. Nevertheless, we would expect a solid to come to equilibrium with the surrounding gas in less than a second, unless the gas is able to penetrate into the interior of the solid.

The experimental data on gas adsorption is collected by placing a solid surface (often in the form of a powder) in an evacuated chamber and heating it for some time, whilst continuing to pump out any evolved gases. When the surface is considered to be 'clean', the sample is taken to the (usually lower) temperature of the experiment. Known amounts of the adsorbing gas are then allowed into the sample chamber and the residual gas pressure is measured after equilibrium has been established. This allows one to estimate the amount adsorbed as a function of the equilibrium pressure.

We will be concerned not only with gas adsorption but also with adsorption from solution, where the situation is very different although the result is often much the same. When considering the adsorption of a solute from solution onto a solid surface, we must recognize that the surface is already covered with the solvent molecules which may interact more or less strongly with it. For systems involving water that interaction is usually very strong. An adsorbate molecule therefore has to compete with a water molecule for a place at the surface and will only be adsorbed if the free energy of the system is lowered by that adsorption. The *rate* at which adsorption equilibrium is attained can be very much slower in this case, partly because it depends on diffusion of the solute to the surface and partly because a competitive reaction or exchange process is usually involved.

6.1.2 *The Langmuir adsorption isotherm*

The amount of material which is adsorbed on a surface, at a particular temperature, depends on the amount of that substance in the gas phase and that dependence is called the **adsorption isotherm**. For surface area determination we are normally concerned mainly with the laying down of the first layer of adsorbate. That is a process which involves interaction between different species. If more than one layer is to be adsorbed, the process involved for the second and subsequent layers is much like condensation of the gas to a liquid. It is normally a physical adsorption process and usually occurs only at pressures close to the normal equilibrium vapour pressure of the liquid. The Langmuir adsorption isotherm gives the relation between the coverage of the first layer, θ, and the gas pressure at a particular temperature.

Langmuir's treatment assumes that all the adsorption sites are equivalent and the ability of the adsorbate to bind there is independent of whether adjacent sites are occupied or not (Atkins 1982). The adsorbed molecules are assumed to be in dynamic equilibrium with the molecules in the surrounding gas:

$$A(g) + M(surface) \rightleftharpoons AM \qquad (6.2)$$

and the rate coefficients for the adsorption and desorption process are k_a and k_d respectively. The rate of adsorption is proportional to the pressure of A, and the number of sites available on the surface, $N(1 - \theta)$, where N is the total number of sites. Therefore:

$$\text{Rate of adsorption} = k_a p_a N(1 - \theta). \qquad (6.3)$$

The rate of desorption is proportional to the amount of gas adsorbed:

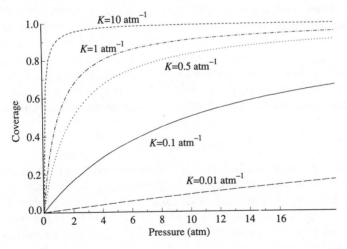

Fig. 6.1. The Langmuir adsorption isotherm (eqn (6.5)) for various values of the constant K.

$$\text{Rate of desorption} = k_d N\theta. \tag{6.4}$$

At equilibrium the rates are equal and by equating eqns (6.3) and (6.4) and solving for θ we arrive at the **Langmuir isotherm**:

$$\theta = Kp_a/(1 + Kp_a) \tag{6.5}$$

where $K = k_a/k_d$. The form of the isotherm for various values of K is shown in Fig. 6.1. Higher values of K mean a higher **affinity** of the gas for the solid.

Note that the coverage increases with pressure and eventually reaches a maximum, corresponding to a monolayer, if K or the pressure is large enough. The quantity K is the equilibrium constant corresponding to the reaction (6.2) and it depends on the temperature. For physical adsorption, lower temperatures favour the adsorption process and so K is expected to increase as the temperature falls. Since K is an equilibrium constant we can use the usual equations of chemical thermodynamics to relate the temperature dependence of K to the enthalpy of adsorption; the appropriate equation is usually referred to as the **van't Hoff isochore** (Atkins 1982):

$$(\partial\ln K/\partial T)_\theta = \Delta H_{ads}/RT^2. \tag{6.6}$$

The subscript θ means that the equilibrium constant at each temperature is measured at constant coverage. Under those conditions, $K = \text{const.} \times (1/p_a)$ and so

$$(\partial\ln p_a/\partial T)_\theta = -\Delta H_{ads}/RT^2. \tag{6.7}$$

(Note the change in sign.) This value of ΔH_{ads} is called the **isosteric heat of adsorption** referring to the fact that it applies to a certain value of the coverage. (Isosteric comes from the Greek meaning 'same space'.) The Langmuir model implies that ΔH_{ads} should be constant but it is more likely to be a function of coverage: at low θ because of the preferential filling of highly active sites and at high θ because of lateral interactions between the adsorbed molecules as they approach close packing on the surface.

The Langmuir isotherm can be tested by rearranging it into a linear form, for example

$$\theta/(1 - \theta) = Kp_a. \tag{6.8}$$

This would be most suitable if we had a direct method of determining the coverage. More usually we have measurements of the volume, v, of gas which is taken up by a given mass of solid, m, at different pressures. Then, if V is the maximum uptake, assumed to correspond to total coverage, we would have $\theta = v/V$ and so, $v/(V - v) = Kp_a$ which can be rearranged to give:

$$p_a/v = p_a/V + 1/KV. \tag{6.9}$$

Thus, if K is independent of coverage, a plot of p_a/v against p_a should be linear with a slope from which the maximum adsorption capacity V/m could be estimated. The intercept would then allow K to be estimated at that temperature. If we know how much area the molecules occupy on the solid surface, the value of V can be used to estimate the surface area of the solid. That method is particularly useful for adsorption from solution where the same equations hold, except that the gas pressure is replaced by the solute concentration (Exercise 6.2.1). Figure 6.2 shows a plot of this sort for the adsorption of CO onto charcoal. Note that the line has a slight curvature which would be expected if K were not strictly constant.

Rather more information can be obtained from studies at different temperatures, but here it is necessary to make measurements at constant coverage in order to make use of eqn (6.7). If we measure the pressure, p_a, required to cause a fixed volume v of gas to be adsorbed at a number of temperatures, the results can be used to evaluate ΔH_{ads} (Exercise 6.1.1). The procedure works well for a physically adsorbed gas but care is needed when dealing with chemisorbed substances, to ensure that equilibrium is reached at each temperature. If the temperature is lowered to values where few molecules have the necessary activation energy, it may take a long time to obtain equilibrium data. This can be checked by determining whether the isotherm is the same going from low to high pressure as it is in the opposite direction.

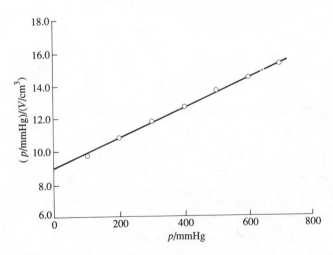

Fig. 6.2. Test of the Langmuir isotherm for CO on charcoal. (From Atkins (1982), with permission.)

6.1.3   *The BET adsorption isotherm*

The Langmuir isotherm is suitable only for situations where the adsorption is limited to a monolayer (and not always even then). When multilayer adsorption is possible, leading ultimately to condensation of the vapour as a liquid onto the solid, a better description is provided by the isotherm developed by Brunauer, Emmett, and Teller, and universally known by their initials. The first layer in that case may sometimes (though rarely) involve chemisorption but all subsequent layers will be physisorbed. A derivation of this equation is given by Atkins (1982) and we will not repeat it here. It is an extension of Langmuir's argument with successive layers being formed independently. An arriving molecule may land on a bare spot or on one on which there is already one or more molecular layers stacked up. The final result, for the volume of gas adsorbed, V, compared with the amount, V_{mon}, required to form a monolayer is:

$$V/V_{mon} = c(p/p^*)/\{(1 - p/p^*)[1 - (1 - c)(p/p^*)]\} \qquad (6.10)$$

where p^* is the equilibrium vapour pressure of the adsorbate and c is a constant related to the adsorption and desorption rate constants.

The coverage as a function of relative pressure is shown in Fig. 6.3. As the value of c increases one can see the clear development of a knee in the curve corresponding to the formation of a monolayer. As the vapour pressure approaches the value of the equilibrium vapour pressure for the condensed liquid, the adsorption increases exponentially and ultimately a macroscopic film of liquid is formed on the surface.

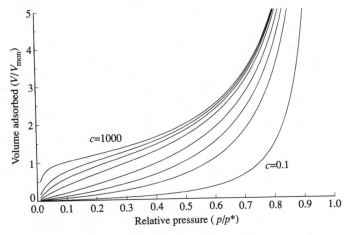

Fig. 6.3. The BET isotherms for various values of $c(= 0.1, 0.5, 1, 2, 5, 10, 100,$ 1000).

Equation (6.10) can be written (Exercise 6.1.2):

$$z/(1 - z)V = 1/cV_{mon} + (c - 1)z/cV_{mon} \qquad (6.11)$$

where $z = p/p^*$. Knowing the vapour pressure of the adsorbate allows one to plot the function $z/(1 - z)V$ as a function of z to establish the validity of the isotherm, and to obtain the values of c and V_{mon} from the intercept and slope. Most systems give a linear plot up to at least $p/p^* = z = 0.3$ (Fig. 6.4). A suitable gas for determining the area of a solid will have a high value of c and in that case the intercept is zero and the slope is $1/V_{mon}$. This allows one to obtain an estimate of the area from a single point measurement (accurate to about 5 per cent) in the relative pressure region 0.1–0.3 (Exercise 6.1.2). The usual gas used is nitrogen at a temperature near to the boiling point ($-196 \,^\circ$C) but krypton is a common alternative, particularly useful for samples of small surface area.

Although for many solids the shape of the adsorption isotherm is very much like that shown in Fig. 6.3, for some value of c the agreement is by no means exact, as is evident from the fact that the data begin to depart from the linear plot in Fig. 6.4 for values of $z(= p/p^*)$ greater than about 0.3. There are a variety of reasons for this sort of 'nonideal' behaviour.

So far we have treated the solid as a smooth uniform surface to which the gas all has ready access. That is not always the case, especially if the solid is finely divided and/or porous. In that case the shape of the isotherm may become very different from that shown in Fig. 6.3, especially for high values of z where the gas is approaching the point at which it can condense into a

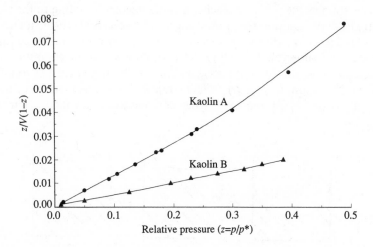

Fig. 6.4. The BET adsorption isotherm for two kaolinite samples, plotted according to eqn (6.10).

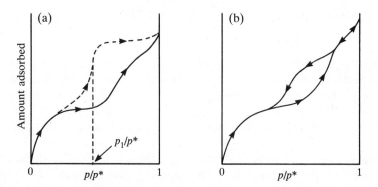

Fig. 6.5. Typical adsorption isotherm for vapour on a porous solid. (a) Full curve, normal behaviour; broken curve, expected behaviour if all pores are uniform in radius. (b) Hysteresis observed when adsorption is followed by desorption. (p^* is the equilibrium vapour pressure at the temperature of the experiment.)

liquid. It may also show some hysteresis effects, that is, the curve may be different depending on whether the pressure is being increased or decreased (Fig. 6.5). One reason for this has already been hinted at in §5.5. As the gas approaches its saturation vapour pressure it becomes possible for condensation to occur in fine capillaries in the solid even before its saturation value is reached, if the resulting radius of curvature of the liquid surface is small enough (Fig. 6.6).

If all the pores had the same radius we would expect the bulk liquid to condense as soon as the pressure reaches a value p_1 corresponding to the radius of curvature of liquid in that pore, as given by eqn (5.37) (broken curve in Fig. 6.5). The curve rises very steeply because every pore can fill almost to the top at the same value of p (Fig. 6.6 (a)). Only when the curvature starts to decrease (Fig. 6.6 (b)) is it necessary for p to be raised further to induce more condensation. Over a very small volume change the meniscus changes from (a) to (b) to (c) and no further condensation can then occur until $p = p^*$.

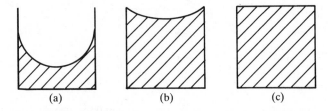

Fig. 6.6. Filling up the pores in a porous solid.

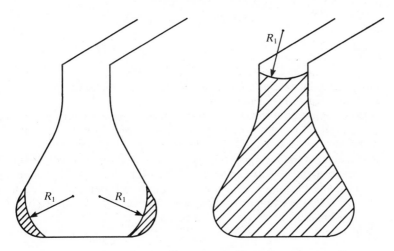

Fig. 6.7. The 'ink-bottle' effect (see text).

In practice, of course, the capillaries are not all of the same size and they therefore fill at different values of p, the smaller ones filling first. The hysteresis shown in Fig. 6.5 can occur in a number of ways of which the most likely is the 'ink-bottle' effect (Fig. 6.7). As p is increased, a point is reached at which the equilibrium curvature is that shown on the left and the pore fills as indicated in (a). If p is increased further the curvature can become flatter and the pore gradually fills up to the point indicated in (b). Now if p is reduced, all of the liquid in the pore can remain in equilibrium with the vapour until the pressure falls back to p_1 again. The desorption curve is therefore much higher than the adsorption curve. It is obvious from the data in Table 5.2 that these effects should not become apparent until the pores are below about 100 nm in radius.

Exercise

6.1.1 Atkins (1982) gives the following data for the pressures of CO required to cause adsorption of 10.0 cm^3 of gas onto a sample (3.022 g) of charcoal. (All volumes corrected to 1 atm pressure and 273 K.) Show from eqn (6.7) that the enthalpy of adsorption can be obtained from a plot of $\ln p$ as a function of $1/T$. What is the value of ΔH_{ads} at this coverage? This is the **isosteric** heat of adsorption.

T(K)	p (mm Hg)
200	30.0
210	37.1
220	45.2
230	54.0

T(K)	p (mm Hg)
240	63.5
250	73.9

Assuming that a CO molecule occupies an area of $0.2\,nm^2$ and that the coverage in this case is 0.4, estimate the area of the solid ($m^2\,g^{-1}$).

6.1.2 (a) Derive eqn (6.11) from (6.10).

(b) Below is given the volume (V in cm^3) of nitrogen gas (corrected to 1 atm pressure and 273 K) adsorbed on a carbon black sample (mass 1 g) as function of pressure (p (mm of Hg)). Plot the volume as a function of pressure. The lowest point of the linear part of the curve is usually identified with monolayer formation. Replot the data in terms of eqn (6.11) assuming that p^* is 760 mm (that is, the measuring temperature is the temperature of boiling liquid nitrogen). Do the estimates of monolayer coverage agree? Take the area of a nitrogen molecule as $0.16\,nm^2$ and estimate the surface area of the sample of carbon black ($m^2\,g^{-1}$).

V	17	22	25	28	30	37	47	54	75
p	15	20	50	80	100	210	320	440	525

6.1.3 Estimate the surface areas of the kaolinite samples in Fig. 6.4 assuming that the mass of solid is 1 g in each case and the volume is measured in cm^3 corrected to 1 atm pressure and 273 K.

6.2 Adsorption at the solid–liquid interface

The methods of measurement of adsorption at the solid–liquid interface are analogous to those used for the solid–gas interface. The clean solid surface, again often in the form of a fine powder, is equilibrated at a fixed temperature, with solutions containing increasing concentrations of a solute and the residual solute concentration, after equilibration, is measured. The difference is assumed to have been adsorbed onto the solid surface (Exercise 6.2.1). Equilibrium can be assumed if the adsorption isotherm is the same as the desorption isotherm.

The adsorption of a coloured molecule, for example, could be followed by measuring the visible absorption spectrum before and after adding the solid. A calibration curve is first prepared, linking the amount of light absorbed at a particular wavelength to the concentration of the solute. The change in light absorption can then be used to estimate the amount of solute adsorbed. Colourless organic adsorbates are often followed by measuring a suitable absorption peak in the UV or IR spectrum. One particularly important example of solid–liquid adsorption in a colloidal system occurs when an oxide or clay mineral is titrated with acid or base. In that case, after each addition of a known amount of acid (or base), the pH is measured to determine how much H^+ or OH^- ion has been taken up by the solid surface. At the same time the pH gives the equilibrium concentration of H^+ or OH^-

remaining in the solution. We will take that matter up in the next chapter.

An argument similar to that used in §6.1.2 can be used to establish the form of the Langmuir isotherm suitable for solutions:

$$n/n_\infty = Kc/(1 + Kc) \tag{6.12}$$

where n is the number of moles of solute adsorbed per unit mass of solid, n_∞ is the number required for monolayer coverage, c is the concentration, and K is a constant (Exercise 6.2.1). Once again the constant K can be treated as an equilibrium constant and its values at different temperatures (evaluated at constant coverage) can be used to determine the (isosteric) heat of adsorption (§6.1.2).

6.2.1 The Freundlich adsorption isotherm

In our derivation of the Langmuir isotherm we assumed that the adsorbate molecules were attached to individual sites on the surface with essentially the same binding energy at each site. At very low pressures (such that $Kp \ll 1$) eqn (6.5) would give simply $\theta = Kp$. That linear relation is obeyed by many but not all adsorbates at low concentration. Sometimes a more appropriate expression to describe the data is

$$\theta = Kp^{1/\nu} \tag{6.13}$$

where ν is an empirical constant (usually lying between 2 and 10). It can be shown (Adamson 1969) that this modified isotherm is to be expected if the

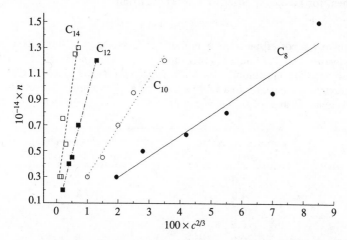

Fig. 6.8. Adsorption of long-chain alkyl sulphates at the oil–water interface in the absence of salts and with coverage less than 0.3. In this case the Freundlich isotherm can be justified theoretically and the exponent $1/\nu$ is expected to be 2/3. (Redrawn from Davies and Rideal (1963, p. 190).)

binding energy varies continuously from site to site on the solid surface. It is known as the **Freundlich adsorption isotherm**, although Freundlich evidently had mixed feelings about it (Davies and Rideal 1963), probably because it doesn't behave properly as p tends to zero. It remains, nevertheless, a useful method of quantifying behaviour in certain systems. It describes well the adsorption of some solutes from solution, where it would take the form: $n/n_\infty = kc^{1/\nu}$. To test for agreement with this equation we would plot $\log n$ against $\log c$ and use the slope to evaluate ν.

The best examples involving the Freundlich isotherm refer to the adsorption of a solute onto the liquid–air interface where it is possible to give some theoretical justification (Davies and Rideal 1963) for various values of the exponent $1/\nu$ ranging from 1/3 to 2/3 (Fig. 6.8). It is, however, often used for describing adsorption data at the solid–solution interface, at least over a restricted range of concentrations. (It is often possible to fit a straight line to a log/log plot between two variables which are not themselves linearly related, and linear plots make interpolation easier.)

Exercises

6.2.1 A sample of 12.5 g of clay mineral is equilibrated with solutions of a cationic surfactant of various concentrations, and the residual concentration of surfactant is determined by UV absorption after equilibrium is established, with the result shown in the table below. Show that if n is the amount (mol g^{-1}) of surfactant adsorbed when the equilibrium concentration is c then, for Langmuir adsorption we should find:

$$1/n = 1/n_\infty + (1/Kn_\infty)c^{-1}$$

where n_∞ corresponds to a monolayer. Assume that the cation is being adsorbed and determine whether the data fit a Langmuir isotherm and, if so, estimate the maximum adsorption capacity for this mass of clay. Hence estimate the surface area of the clay (m^2 g^{-1}) assuming that the surfactant occupies an area of 0.27 nm^2 per ion.

mmoles added to sample	mmoles remaining in solution
2.0	1.83
4.0	3.70
6.0	5.56
8.0	7.47
10.0	9.41
12.0	11.35
16.0	15.25
20.0	19.20

6.3 Adsorption at the liquid–air interface

It should be obvious that the general method of studying the solid–gas interface is not applicable to the study of the liquid–gas interface. Nor is there much interest in studying the adsorption of a gas at the surface of a liquid, since once it is adsorbed it will diffuse into the body of the liquid with little hindrance. Much more interesting is the study of the adsorption of a solute from the bulk liquid into the liquid–gas interface. The liquid–gas interface is also a lot simpler in many respects than the solid–gas interface. For one thing, it is molecularly smooth, and we already know that it has a well defined surface energy (§5.2.1). This surface energy can, in fact, be used to study the amount of adsorption occurring because adsorption of a solute will occur *only if it leads to a lowering of the interfacial energy*.

The extent to which γ is lowered is a direct measure of how much is adsorbed and the exact relation between the two was first derived by J. Willard Gibbs in his famous series of papers on the equilibrium of heterogeneous substances (Gibbs 1875–7). To understand his analysis we must extend the notion of **surface excess** quantities introduced in §5.3.2 to the case of a solute. We didn't need this in Chapter 5 because it dealt only with systems containing a single component.

In a system containing an interface between two phases and more than one component, (say a solution of an organic compound in water in equilibrium with the vapour) we find that the presence of the interface influences the distribution of the solute in the water. The solute may tend to accumulate in the interface at a concentration higher than would be expected from its bulk concentration (positive adsorption) or it may be driven away from the interface (negative adsorption). Whether it accumulates or not is determined by how it affects the interfacial energy or surface tension.

We will deal with a system made up of just two components, of which one can be treated as the solvent and the other as the solute. The solution is assumed to be in equilibrium with its vapour which will, in general, consist of both components, although the solvent vapour will often predominate. The volumes of the phases are again V' and V'' and the concentrations of the solvent are c_1' and c_1'' in the two phases, whilst the solute has concentrations c' and c''. (It may seem strange to talk of the concentration of the solvent but in this analysis we must account for all of the material. In the case of a solution of ethanol in water, for example, there is no real distinction between the components as to which is solvent and which solute, and their respective concentrations can be specified in terms of the number of moles of each in a given total volume.)

For each component we can write, for the number of moles present in each phase:

$$n' = c'V'; n'' = c''V''; n_1'' = c_1''V'; \text{ and } n_1'' = c_1''V''. \quad (6.14)$$

Once again, the extra amount of each component that can be accommodated in the system because of the presence of the interface is given by

$$n^\sigma = n - c'V' - c''V'' \text{ and } n_1^\sigma = n_1 - c_1'V' - c_1''V'' \quad (6.15)$$

where n and n_1 are the total numbers of moles of solute and solvent present in the system respectively. Note that all of the quantities on the right-hand side of these equations are quite well defined. A **surface concentration** (moles per unit area) can be defined as: $n^\sigma/A = \Gamma$ for the solute and $n_1^\sigma/A = \Gamma_1$ for the solvent. (For a multicomponent system the surface concentration for each of the components is defined in the same way: $\Gamma_i = n_i^\sigma/A$ where A is the area of the interface.)

You will recall from §5.3.2 that we divided the system up into the volumes V' and V'' with an infinitesimally thin surface and assumed that a sharp break occurred there, with the density, and possibly the pressure, changing discontinuously at that surface. When we come to consider the concentrations of solvent and solute, the consequences of that model become very important because it turns out that the measured surface excess of solute

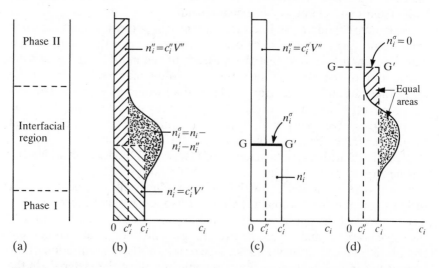

Fig. 6.9. The concentration of a component i across an interface can be plotted and will generally show rapid changes through the interfacial region (b). The Gibbs model assumes that the concentrations of the components are constant and equal to their bulk values (cross-hatched regions in (b)) right up to some arbitrary plane. The excess material (dotted region in (b)) is assumed to be adsorbed at that plane GG' in (c). Moving the plane to a new position (d) can drastically alter the magnitude of the surface excess. (See text for details.)

depends critically on where the surface is placed. Figure 6.9 shows how this comes about.

Figure 6.9(a) shows the 'real' situation with a finite region where the concentration of solute is varying rapidly from its constant value (c') in phase $'$ to its value in phase $''$ with a possible higher concentration (positive adsorption) in the interface. The Gibbs convention requires that the concentrations within the two phases be regarded as constant right up to the dividing surface (Fig. 6.9(b)). That leaves all the material in the dotted region as the surface excess and all of it is assumed to be adsorbed in the plane GG$'$. Notice that if we move the dividing surface upwards we can make n^σ smaller. We could make it zero *or even negative* just by moving the surface a short distance, possibly as little as a few nanometres.

How can such a number be of any use then? How can we make a sensible choice for the position of this plane if it must be done so exactly? The answer is very simple: we normally choose the position of the dividing surface exactly at the place which will make the surface excess *of the solvent* (n_1^σ) equal to zero. Once that position is chosen it must be used for the calculation of all of the surface excess quantities. We may never know precisely where it is in physical terms but that is unimportant. The resulting value for the surface excess of the solute is a perfectly well defined quantity as we shall see shortly. It corresponds to the *excess in concentration of the solute at the interface relative to the concentration of the solvent* (§6.3.2).

6.3.1 The Gibbs adsorption isotherm

The Gibbs isotherm is derived from a purely thermodynamic argument. It doesn't tell us how much is adsorbed as a function of solute concentration like eqn (6.12). Rather it tells us how to estimate the amount adsorbed if we can measure the surface tension γ and how that depends on solute concentration. It is, in effect, a variation of the Gibbs–Duhem equation (5.31) and is derived in a similar way.

We begin with eqn (5.15):

$$dU^\sigma = TdS^\sigma + \gamma dA + \sum(\mu_i^\sigma dn_i^\sigma) \qquad [5.15]$$

and imagine that this change in U^σ is brought about by an increase $d\chi$ in the size of the surface, so that all of the extensive surface properties (U, S, A and n_i) are changed in the same proportion:

$$dU^\sigma = U^\sigma d\chi; dS^\sigma = S^\sigma d\chi; dA = A d\chi; \text{ and } dn_i^\sigma = n_i^\sigma d\chi. \quad (6.16)$$

Substituting these values into (5.15) gives:

$$U^\sigma d\chi = TS^\sigma d\chi + \gamma A d\chi + \sum(\mu_i^\sigma n_i d\chi). \qquad (6.17)$$

This equation can now be integrated from 0 to 1 (which corresponds to doubling the size of the surface whilst keeping all the properties the same). The result is

$$U^\sigma = TS^\sigma + \gamma A + \sum (\mu_i n_i). \qquad (6.18)$$

(We can drop the superscript σ on μ_i because we proved in §5.3.3 that the chemical potential was constant throughout the whole system.) This is an important general equation for the surface energy. It can be differentiated to give

$$\mathrm{d}U^\sigma = T\mathrm{d}S^\sigma + S^\sigma \mathrm{d}T + \gamma \mathrm{d}A + A\,\mathrm{d}\gamma + \sum(\mu_i \mathrm{d}n_i + n_i \mathrm{d}\mu_i) \quad (6.19)$$

and comparing this with eqn (5.15) shows that

$$0 = S^\sigma \mathrm{d}T + A\,\mathrm{d}\gamma + \sum n_i^\sigma \mathrm{d}\mu_i. \qquad (6.20)$$

This is the same trick that was used to derive the Gibbs–Duhem equation (5.31) but we get a different term this time, because the system has an extra degree of freedom from the surface effect but loses one from the fact that the surface phase has zero volume. Again eqn (6.20) spells out the limits of variation in the system and this time we use it to extract the amount of surface adsorption.†

At constant temperature $(dT = 0)$:

$$-\mathrm{d}\gamma = \sum_i \frac{n_i^\sigma}{A}\,\mathrm{d}\mu_i = \sum_i \Gamma_i \mathrm{d}\mu_i. \qquad (6.21)$$

This is called the **Gibbs adsorption isotherm**. It doesn't look much like the earlier expressions but it can be made a little clearer by treating the special case of a two-component system, in which the Gibbs dividing surface is placed so that n_i and hence Γ for the solvent is zero. Then we have

$$-\mathrm{d}\gamma = \Gamma_{\text{solvent}}\,\mathrm{d}\mu_{\text{solvent}} + \Gamma_{\text{solute}}\,\mathrm{d}\mu_{\text{solute}}$$

$$= \Gamma\mathrm{d}\mu$$

or

$$\Gamma = -\mathrm{d}\gamma/\mathrm{d}\mu \text{ at constant temperature} \qquad (6.22)$$

where Γ and μ refer to the solute. Then writing $\mu = \mu^0 + RT\ln c$ we get, for the surface excess

---

†The Gibbs–Duhem equation is usually used to calculate the chemical potential of one component (for example, a solute) from the value of another (the solvent) which is more easily accessible.

$$\Gamma = -(1/RT)(d\gamma/d\ln c) = -(c/RT)d\gamma/dc. \qquad (6.23)$$

Thus, the excess is not given directly in terms of the solute concentration, but rather in terms of the effect of the solute on the surface tension. If the solute is positively adsorbed it will lower γ and, for dilute solutions, we will expect the lowering to be proportional to the concentration:

$$\gamma_0 - \gamma = KC$$

and so $d\gamma/dc = -K$ and, from eqn (6.23):

$$\Gamma \text{ (moles per unit area)} = (K/RT)c \qquad (6.24)$$

which is of the same form as the Langmuir isotherm at low concentrations.

A typical example of the use of the Gibbs adsorption isotherm is given in Fig. 6.10. It shows the surface tension as a function of concentration for a number of aliphatic alcohols from hexanol to decanol. These are not very soluble in water and the solubility decreases as the chain length increases. They all cause γ to decrease, so they are all positively adsorbed at the water–air interface. The amount adsorbed obviously increases as the concentration increases because the slope of the curve of γ against log c becomes more negative as c increases. There are, however, signs that the surface is becoming saturated at the higher concentrations (Exercise 6.3.1).

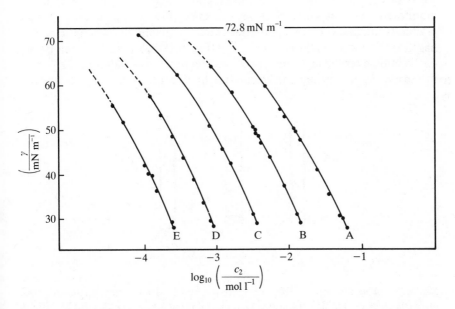

Fig. 6.10. Surface tensions of aqueous solutions of straight chain aliphatic alcohols. A–E: C_6–C_{10} primary alkanol.

6.3.2 Measuring surface tension

If we are to use γ as a measure of adsorption at the water surface we need
to be able to measure it accurately, and a wide variety of methods have been
developed for the purpose. This is not the place to describe in detail how the
measurement is done, because accurate measurement requires a good deal of
careful preparation. The principle behind the main methods can, however,
be fairly readily understood from the ideas that we have already introduced.

To begin with, we can measure the height to which a solution rises in a
capillary. Solutions of an alcohol in water at different concentrations will
rise to different heights depending on their γ values and densities:

$$h = (2\gamma \cos \theta)/r\rho_1 g \qquad (5.43)$$

so for liquids which all wet the capillary to the same extent we have

$$\gamma_1/\gamma_2 = h_1\rho_1/(h_2\rho_2). \qquad (6.25)$$

Using the value for pure water as a standard, the results could be used to construct a curve like those shown in Fig. 6.10.

Another procedure is to measure the pressure required to blow a bubble
in the liquid. A capillary tube is held under the surface of the solution and
gas is slowly blown into it while the pressure of the gas stream is carefully
monitored. The pressure rises as the curvature of the bubble surface
gradually increases (Fig. 6.11) until it reaches a maximum, when the bubble
has the same radius as the capillary tube (Fig. 6.11(b)). Further development
of the bubble results in a decrease of the pressure as the radius of the bubble
increases. By measuring the **maximum bubble pressure** using a capillary

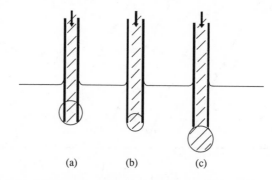

(a)   (b)   (c)

Fig. 6.11. The maximum bubble pressure method of determining surface tension.
(a) Early stage of bubble formation. Low curvature and low pressure. (b) Maximum
pressure is required when the bubble has a radius equal to that of the capillary. (c)
As the bubble grows bigger, the pressure required to increase its size becomes smaller.

of known radius, the surface tension can be calculated from the Young–
Laplace equation:

$$\Delta p = 2\gamma/r \qquad (5.6)$$

(Note that a bubble under water has only one surface so the formula is the
same as for a drop of liquid in air.) In practice it is again preferable to use
clean water as the standard and to measure the bubble pressures of the
solutions with respect to that so that any uncertainties in the radius of the
capillary are eliminated.

The shape of a drop, as it hangs from the end of a capillary, is determined
by the balance between the surface tension forces (which tend to hold it as
a sphere) and the force of gravity which makes it sag to some extent
(Fig. 6.12(b)). A careful study of the drop profile (usually using a photo-
graphic procedure) can, therefore, give the surface tension. The mathematics
involved in describing the profile is not easy but the method has been in use
since late last century. It is particularly useful when ageing effects are being
studied because the profile can be studied as a function of time.

One of the most widely used and effective methods of determining γ is to
measure the force required to pull a plate or a ring from the surface. The surface tension force acts around the perimeter of the object which is being
withdrawn so the force is easily calculated as force = $\gamma \times$ perimeter. In the
Du Noüy ring method, that force is measured directly using a microbalance
to determine the apparent weight of a ring of platinum as it is pulled through
the water/air surface (Fig. 6.13). The meniscus forms on both sides of the
ring so the force is given by $f = 4\pi r\gamma = mg$ where m is the apparent mass
as the ring comes through the surface.

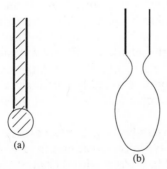

(a)

(b)

Fig. 6.12. The shape of a hanging drop. (a) In the absence of gravity. This would
be the case if the surface tension force greatly exceeded the gravitational force (as
happens with a mercury drop suspended in water), or where the density difference
between the drop and the liquid in which it is being formed is very small. (b) The
sagging which occurs when the gravitational forces are significant.

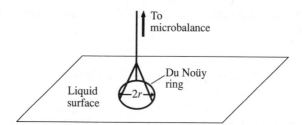

Fig. 6.13.   The principle of the Du Noüy tensiometer.

   A very similar method is used with the **Wilhelmy plate** in which a rectangular piece of mica dangles in the surface. This is the technique used in
the Langmuir trough which has been widely used to study the adsorption of
soluble, and more particularly insoluble, surface active agents at the
air–water and oil–water interface. A detailed study of that work would take
us too far afield but some of the basic concepts are taken up in the next
section.

6.3.3   Adsorbed films

In an earlier age, before the advent of television and video games, our
forebears made do with much simpler toys. One of the more intriguing from
our point of view was a toy boat (very light in weight and only a few centimetres in size) which could be placed on a clean water surface in a basin
where it would happily buzz around for hours on end, powered only by a
small piece of an organic compound called camphor wax stuck onto the rear
(Fig. 6.14). The camphor slowly spread out from the solid to cover the
surface of the water and as it spread behind the boat it lowered the surface
tension there. The higher tension in front of the boat would pull it forward.
Because camphor is fairly volatile, the film of camphor evaporates from the
surface and, provided the area of the basin is large enough, the water surface
in front of the boat continues to be clean enough to allow the boat to keep
moving. The boat can go on moving until all of the camphor has spread and
evaporated.
   This 'toy' illustrates a number of important aspects of the surface tension.
The spreading process itself occurs because the free energy of the camphor
is lowered as it spreads to interact with the water surface. The lowering of
the surface tension can be made to do work in a very obvious sort of way.
The still further lowering of free energy as the camphor evaporates from the
large water surface occurs because of the even lower free energy of the
camphor in the gas phase, compared to its value on the water surface
(Exercise 6.3.2).

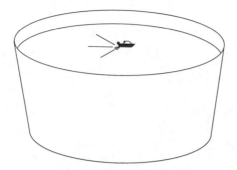

Fig. 6.14. The camphor boat.

The change in the surface tension ($\gamma_{\text{clean}} - \gamma_{\text{camphor covered}}$) can be inter-preted as a (two-dimensional) pressure which pushes the boat forward. That **spreading pressure**, as it is called, is just one aspect of the notion of the **surface pressure** (π) of an adsorbed film at the air–water interface:

$$\pi = \gamma_0 - \gamma \qquad (6.26)$$

where γ_0 is the surface tension of the clean liquid surface and γ is the value with an adsorbed (or spread) film.

The camphor boat works well partly because camphor is essentially insolu-ble in water and it spreads easily from the solid to cover the water surface. Other organic compounds can spread very easily from a droplet to form a film. (Exercise 6.3.3). In order to investigate such systems more carefully, however, it is better to dissolve the organic compound in a suitable (volatile) solvent and add a known volume of the solution to the surface of the water. The solvent then evaporates and the insoluble film remains, and knowing the initial solution concentration and the volume added we know exactly how much material has been placed on the surface. If the available area per molecule is very large, the molecules of the film will behave as though they are in the form of a gas. If the area is small, so that the molecules of the film are almost touching one another, then the film will behave as a liquid, or even a 'solid'.

How can we distinguish between a gas and a liquid or solid when all we have is this minute amount of material in the surface? (The amount of camphor required to cover a square metre of water with a thickness of one molecular layer is only about 1 mg.) The solution to this problem was pro-vided by the same Irving Langmuir who did so much pioneering work in the study of adsorption on solid surfaces. A crude representation of his apparatus is shown in Fig. 6.15.

There are two methods in general use for determining the value of π as a function of the area available per molecule. One is the Wilhelmy plate

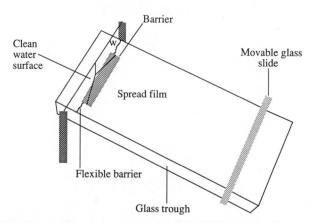

Fig. 6.15. The Langmuir trough or surface balance. The trough is filled to the brim with water and the surface is cleaned by moving the glass slide along the surface. This pushes any impurities to one end where they can be sucked away with a capillary tube attached to a vacuum line. The film is spread on one side of the barrier and the pressure it exerts is measured by the twisting movement produced in a torsion wire (w). The area available for the molecules of the film can be varied by moving the glass slide. Note the flexible barrier (usually a waxed silk thread) which separates the clean water surface from the film surface.

method and the other depends on measuring directly the pressure which the film produces against a barrier floating in the surface. The trough in Fig. 6.15 uses the barrier method which works best for insoluble films. The film is spread on one side of the barrier whilst the other side has a clean water surface.

In the Wilhelmy plate method, a sheet of mica dangles vertically in the surface of the liquid in the trough. The value of π is determined by measuring the force which must be applied to the plate to hold it at a fixed depth in the liquid. As the surface tension changes, the weight (mg) of the meniscus hanging from the plate also changes and this is what is, in effect, being measured:

$$f = mg = 2(a + b)\gamma \tag{6.27}$$

where a and b are the length and thickness of the plate.

(The weight of the plate itself, corrected for bouyancy, is also involved but remains constant if the depth of immersion is fixed (Fig. 6.16).) The contact angle must also be assumed to remain fixed; it must be known, and the method really only works when it can be assumed to be zero. This method is not usually as sensitive as the barrier method but must be used if the substance being studied is able to dissolve in the water. The barrier method doesn't work in that case because one cannot maintain a clean water surface behind the barrier.

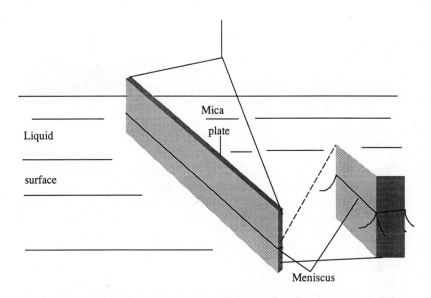

Fig. 6.16. Operation of the Wilhelmy plate.

Whichever method is used to determine the surface pressure, an experiment is done by depositing a known amount of the substance to be studied (usually in the form of a solution in a volatile solvent) onto the cleaned water surface. One then measures the pressure as a function of the area available to the substance. That area can be reduced by moving the waxed glass slide over the water surface (Fig. 6.15).

When the film is highly dilute, so that the molecules are relatively far apart, the film behaves like a gas, *but it is a two-dimensional gas*. In the limit of extremely low pressure ($\pi \to 0$) where the area per molecule, A, is very large, we observe an analogue of the perfect gas law:

$$\pi A = kT. \tag{6.28}$$

Note that the area per molecule here takes the place of the volume per molecule for an ordinary three-dimensional gas. The two-dimensional pressure, π, is sometimes viewed as a sort of osmotic pressure of the film (Fig. 6.17) with the barrier being regarded as a sort of semipermeable membrane (that is, permeable to the water but not to the film molecules).

The ideal behaviour suggested in eqn (6.28) is rarely observed because the effective pressure in these monolayers is much higher than one might expect. Suppose the surface film reduces the value of γ from its clean water value to, say, 70 so that $\pi = \gamma_0 - \gamma = 72-70 = 2\,\mathrm{mN\,m^{-1}}$, and we assume a

Adsorption at interfaces

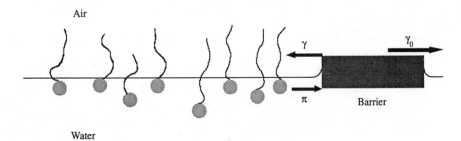

Air

γ γ_0

π   Barrier

Water

Fig. 6.17. The balance of forces on the barrier separating a clean water surface from the film covered surface. The magnitude of the pressure, π, depends on the number of molecules bombarding the barrier from the left. In this sense it is an osmotic type pressure with the barrier acting as a semipermeable membrane. Alternatively one can consider the difference in the tensions acting on both sides of the barrier.

thickness, t, for the monolayer of 1 nm then we might interpret π as a three-dimensional pressure, $p = \pi/t = 0.002/10^{-9} = 2 \times 10^6 \, \text{Nm}^{-2} \approx 20 \, \text{atm}$. It is hardly surprising that at such pressures the monolayer is no longer behaving as an ideal gas at room temperature. A good representation can usually be given by the equation

$$\pi(A - A_0) = xkT \qquad (6.29)$$

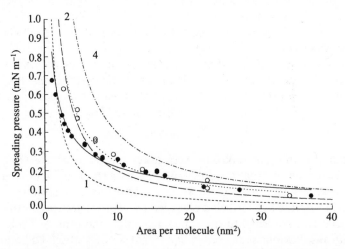

Fig. 6.18. Some typical force–area curves for long-chain organic esters. Vapour and gaseous films. Eqn (6.29) is illustrated for $x = 1$, 2 and 4. Data points are for myristate (●) and laurate (○) with best fit curves using a simple power law relation.

where A_0 would be expected to bear some relation to the area occupied by the molecules when they are close packed, and x is an empirical correction factor. This is very similar to the behaviour of some real gases like hydrogen (which obeys the Amagat equation $[P(V - nb) = nRT]$). Figure 6.18 shows some typical π–A curves for long chain organic esters adsorbed at the air–water interface; they are among the closest to ideal of all spread monolayers.

When the pressure is increased further, or the area available per molecule is reduced, the gaseous film can condense to form other two-dimensional phases. Langmuir distinguished a liquid-expanded state in which the area per molecule was less than about $0.5 \, \text{nm}^2$, a liquid-condensed state, and, finally, a solid state where the area per molecule was only about $0.2 \, \text{nm}^2$.

The molecules can fit into such small areas only if they are lined up at the surface with their minimum cross-section in contact as indicated in Fig. 6.19. Further compression of the film results in its collapse as the molecules are physically pushed out of the surface layer.

6.3.4 Applications

We need an understanding of the behaviour of surfactant molecules at the air–water interface in order to understand how they are able to stabilize a soap film or assist in the flotation of a mineral particle. The Langmuir trough has also been used to examine the interaction between protein molecules and surfactants like lecithin which make up an important part of

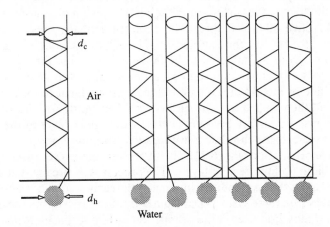

Fig. 6.19. A typical surface-active molecule adsorbed at the air–water interface. When the molecules become packed together so that they are effectively touching, the surface can accommodate no more material. This may be limited by the diameter of the head groups, d_h, or that of the chains, d_c.

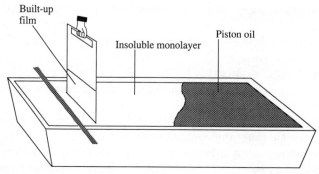

Fig. 6.20. Deposition of Langmuir–Blodgett films using a piston oil to force the adsorbate onto the underlying surface.

the biological cell membrane. Those interactions are very complex and subtle and much of our understanding of them is based on this sort of work. Given the concept of a monolayer as a two-dimensional phase which can exhibit the properties of a liquid or a liquid in equilibrium with its (two-dimensional) vapour or a gas, we can understand the cleaning action of detergents or the dispersing action of a surfactant. We will be discussing those ideas in Chapter 10.

One direct application of the adsorbed film which has created a great deal of interest recently is the production of **Langmuir–Blodgett films**. These are multilayer films of oriented molecules on the surface of a solid, and they are produced using the arrangement shown in Fig. 6.20. The piston oil forces the spread film to coat the surface of the solid. The piston oil is a surface-active agent like olive oil which spreads to produce a steady pressure of about $33 \, \text{mN m}^{-1}$ (rather like the camphor on the back of the boat in Fig. 6.14). It is separated from the material to be deposited by a silk thread with some paraffin jelly (Vaseline) on it to make it nonwetting. The solid to be coated (a glass slide or metal strip) is first treated with an agent to make it hydrophobic and then dipped into the surface and then gently withdrawn. As it is pushed down, its surface becomes coated with a single layer of the material A on the surface. As the slide is withdrawn, a second layer is deposited in the reverse orientation (Fig. 6.21). By immersing and withdrawing the slide, one can build up hundreds of layers. The process is so precise that the films were used as thickness standards in some of the early studies of radioactivity.

The interest in these systems now comes from the area of microelectronics where they may find uses in the production of high-capacity memory chips and other microminiature components for computers. Experiments are also being conducted on the conversion of solar energy into electricity using films containing substances which can absorb a photon and use the energy to release an electron as part of a photochemical cell.

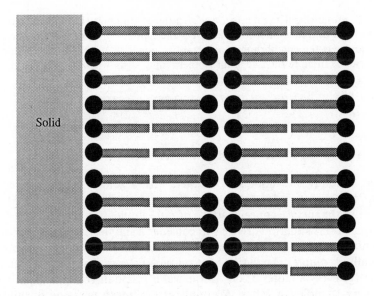

Fig. 6.21. Structure of a Langmuir–Blodgett film on a solid surface. The first layer renders the surface hydrophobic and subsequent layers produce alternately hydrophilic and hydrophobic surfaces.

Exercise

6.3.1 Estimate from Fig. 6.10 the amount of adsorption as a function of concentration for the octanol ($C_8H_{17}OH$) solution assuming that the temperature is 25 °C. Estimate the surface concentration when saturation is reached. Assuming that this corresponds to a monolayer, estimate the cross-sectional area of the molecule.

6.3.2 Camphor is a C_{10} cyclic hydrocarbon with a single ketone group, with which it interacts with water. Discuss the processes involved in the operation of the camphor boat in terms of the enthalpy, entropy, and free energy changes occurring at each stage in the process.

6.3.3 A simple demonstration of the spreading of a monolayer from a droplet is provided by an experiment which was first used in the last century to estimate the size of molecules (which at that stage were regarded by many scientists as mythical objects). Water is placed in a basin (Fig. 6.14) and the surface sprinkled with talcum powder. A single drop of a 0.1 per cent solution of oleic acid (*cis* 9-octadecenoic acid) in ether is gently deposited in the middle of the surface. We observe the talcum powder to be pushed away from the drop to form a clear ring of radius 10 cm. Assuming that the ether evaporates to leave a monolayer of oleic acid, estimate the area occupied per molecule assuming that the droplet had a mass of 0.020 g.

6.3.4 The following data refer to the adsorption of un-ionized butyric acid molecules at the benzene–water interface. n is the number of moles per cm^2 adsorbed and c is the equilibrium concentration (mmol l^{-1}). Determine whether they can be fitted to the Langmuir isotherm and, if so, what the area per molecule is when monolayer coverage is attained.

$10^{10}n$	0.76	1.77	2.78	3.78	4.04	4.54
c(mM)	13.5	37	100	214	355	503

6.4 Adsorption at the liquid–liquid interface

When there are two liquids in contact, with a solute distributed between them, the surface adsorption or surface excess must be evaluated with respect to the surface concentrations of *both* liquids. To see what the consequences of this situation are we will consider a simple 'thought experiment'. A solute is distributed between a water and an oil phase (assumed to be mutually insoluble). Analysis of the oil phase shows that there is 1 mole of solute for every 4 moles of oil. In the water phase there is 1 mole of solute for every 10 moles of water. A sample of the interface region of thickness 0.1 mm and volume 10 cm^3 (taking with it some of each of the bulk phases) gives a mixture of water, oil, and solute in the molar ratio 5:4:2 containing exactly 0.200 moles of water. What is the surface excess of the solute? *Solution*: The interface region sampled has an area of $10/0.01 = 1000 \, cm^2$, and associated with that area is 0.2 moles of water, $0.2 \times 4/5 = 0.16$ moles of oil and $0.2 \times 2/5 = 0.08$ moles of solute. If the bulk concentration of solute in the water phase stretched right up to the interface there would be $0.2/10 = 0.02$ moles of solute in our sample coming from the water phase. If the oil behaved in the same way there would be $0.16/4 = 0.04$ moles of solute coming from that source. The *excess* of solute present over what would be expected if each phase had the bulk composition right up to the dividing surface is therefore, $0.08 - 0.02 - 0.04 = 0.02$ moles. The surface excess concentration is therefore 0.02 mole/1000 $cm^2 = 0.2$ moles per m^2.

We should point out that such an enormous amount of material could not be confined to a plane. The area occupied by a molecule is normally at least 0.15 nm^2 and so 0.2 moles, if present as a close-packed monolayer, would occupy an area of $0.2 \times 6 \times 10^{23} \times 0.15 \times 10^{-18} = 18\,000 \, m^2$. That will give you some idea of why it is difficult to make these adsorption measurements directly. A partial monolayer of the solute would show a concentration difference about 20 000 times smaller than the one we postulated here. That would be very difficult to analyse for by normal methods. This is why the Gibbs adsorption isotherm is so useful. It allows us to estimate the excess simply by measuring the change in γ with solute concentration (see Exercise 6.3.1).

The interface between an oil and water phase can be studied using modifications of the apparatus and techniques used at the air–water interface. Again the results are of great interest in coming to an understanding of the way in which molecules are arranged in the membranes around the biological cell. Although the cell membrane is much more complex and structured than the simple arrangements we can study using the Langmuir trough it is, nevertheless, the ideas obtained from these studies which form the basis of our understanding of those more complicated biological systems.

References

Atkins, P. W. (1982). *Physical chemistry*, (2nd edn). Oxford University Press.
Adamson, A. W. (1969). *Physical chemistry of surfaces*, (3rd edn), p. 410. Interscience, New York.
Davies, J. T. and Rideal, E. K. (1963). *Interfacial phenomena*, (2nd edn), p. 185. Academic, New York.
Gibbs, J. W. (1875-7). On the equilibrium of heterogeneous substances. *Transactions of the Connecticut Academy*, **III**. (Reprinted (1961) as *The Scientific Papers of J. Willard Gibbs*, Vol. 1. Dover, New York.)

ELECTRICALLY CHARGED INTERFACES

7.1 The mercury–solution interface

We noted in §2.3 that many important properties of colloidal systems were influenced by the electrical charges on the particle surface. When it is immersed in an electrolyte solution, a charged colloidal particle will be surrounded by ions of opposite sign so that, from a distance, it appears to be electrically neutral. The surrounding ions are, however, able to move under the influence of thermal diffusion so that the region of charge imbalance, due to the presence of the particle, can be quite significant relative to the size of the particle itself. For very small particles (c. 50 nm) the disturbance it creates can stretch out to several particle diameters. The arrangement of electric charges on the particle, together with the balancing charge in the solution, is called the electrical **double layer**, and it has been studied on various surfaces for well over a century.

A closely related situation occurs when a metal surface is placed in an electrolyte solution, to form the plate of an electrochemical cell. Indeed, it was the study of electrochemical cells which first gave us an understanding of what happens at an electrically charged interface. One of the most widely studied electrode surfaces, and the one which has provided some of the most reliable information, is the interface between a mercury drop and an electrolyte solution.

A small drop of mercury issuing from a glass capillary under the surface of an electrolyte is probably the closest we can get to an ideal system for study (Fig. 7.1). The mercury is contained in a reservoir, M, and as it drops from the lower end of the capillary, C, a new (and clean) surface is continually being created. The surface is also molecularly smooth, which makes the results easier to interpret. Since it is a liquid–liquid interface it also gives us the opportunity to measure the interfacial tension and, as we shall see,

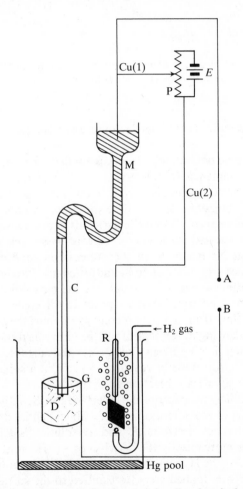

Fig. 7.1. Schematic arrangement for determining the electrical capacitance across the surface of a mercury drop, D (see text). G is an electrode made of, say, Pt gauze and R is an H_2/Pt reference electrode.

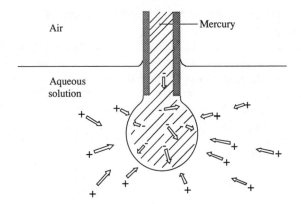

Fig. 7.2. Movement of charge as a result of applying a potential which makes the mercury more negative with respect to the solution.

that provides a wealth of information about the adsorption processes at the interface.

When an electrical potential difference is applied across an electrode interface, there is a tendency for a current to flow in one direction or the other. Electrochemists distinguish two different types of electrode: **perfectly polarizable** and **reversible**. In practice, these are extremes to which real electrodes can only approach, but the basic difference between them is that, in the perfectly polarized electrode, no current passes across the interface. For the reversible electrode, when the potential is applied, current flows across the interface until the system has adjusted itself to the applied potential. In the experiments we will discuss, the mercury–solution interface is regarded as perfectly polarized. When a potential difference is applied across it, electrons will flow from the bulk mercury to its surface or vice versa and there will be a balancing movement of charge in the surrounding electrolyte (Fig. 7.2), but there is no current flow across the surface. (The activation energy for the electron transfer reaction to or from a suitable ion† is just too large for any significant reaction to occur.)

The potential difference across the interface can be adjusted simply by altering the setting of the potentiometer, P. The potential drop between the interior of the reference electrode, R, and the solution is determined by the concentration (or more exactly the activity) of the H^+ ions in the solution (and the pressure of the H_2 gas). Any change dE in the setting of the potentiometer is, therefore, immediately transmitted to the surface of the drop.

† In other words there are no easily oxidizable or reducible ions in the solution. The electrolyte will usually be a simple alkali metal salt like NaCl or KNO_3.

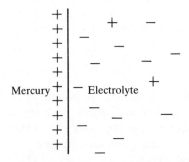

Fig. 7.3. Arrangement of charge at the mercury–solution interface.

Since no current can flow through the drop surface, the only effect is the gradual build-up of a charge on the drop, with a counterbalancing charge in the surrounding medium. The double-layer charge (Fig. 7.3) behaves like an electrical capacitor and the magnitude of the capacitance can be measured with a suitable instrument at terminals AB (Fig. 7.1).

The system shown in Fig. 7.1 could be used to study the behaviour of, say, $HClO_4$ or HNO_3 solutions, since it turns out that the H^+ ion is not very easy to reduce on the mercury–solution interface (that is, negligible current flows through the interface so long as the potential is not made too negative with respect to the reference electrode). Typical experiments of this kind take two forms. Either the bulk activity of the acid is kept constant and the potentiometer setting, E, is changed, or E is kept fixed and the activity (or concentration) of the acid is changed. In either case we want to know how the surface of the mercury is affected and that can be done by following the interfacial tension, γ. We will find that such studies provide many important clues to the understanding of the behaviour of electrically charged colloidal systems.

We also find this to be a very important area for the application of classical thermodynamics, which allows us to manipulate the data, and extract the information we need, without introducing any new assumptions. To do so we must first introduce another modification of the chemical potential, similar to that used in eqn (2.2). The **electrochemical potential**, $\bar{\mu}$, of an ion of valency z when it is at an electrostatic potential of ϕ is defined as follows:

$$\bar{\mu} = \mu + ze\phi \tag{7.1}$$

where μ is its chemical potential and e is the proton charge. (Note that z carries its sign with it.) Once again, it can be shown that to establish equilibrium in a system containing electrical fields, where the electrical potential varies from place to place, it is not the chemical potential but rather the electrochemical potential of each of the charged species which must be constant throughout the system. For a charged interface, we then find that the appropriate form of the Gibbs adsorption isotherm (eqn (6.21)) is

$$-d\gamma = \sum_i \Gamma_i d\bar{\mu}_i. \tag{7.2}$$

The components whose electrochemical potentials are involved in the summation must include the electrons, which form the electrical charge, σ_0, on the metal surface. It can be shown (Hunter 1987) that for the simple situation we are considering, of an electrode dipping into a solution of a salt (or an acid in our case), eqn (7.2) can be rearranged to give us a more immediately useful form:

$$-d\gamma = \sigma_0 dE + \Gamma_- d\mu_{salt} + \Gamma_{water} d\mu_{water}$$
$$= \sigma_0 dE + \Gamma_- d\mu_{salt} \tag{7.3}$$

where σ_0 is the surface charge density on the metal electrode and Γ_- is the **relative** surface excess of the anion taking Γ_{water} as zero (Fig. 6.9). We measure the surface excess of the anion in this case because the reference electrode is *reversible to the cation* (H^+ in this case). If the reference electrode happens to be reversible to the anion, one obtains Γ_+. Fortunately we don't need to measure both because the electro-neutrality condition requires that the charges on either side of the interface must balance one another. They are thus linked by the equation

$$F(z_+\Gamma_+ + z_-\Gamma_-) = -\sigma_0 \tag{7.4}$$

where F is the faraday of charge (96 485 coulombs per mole).

Equation (7.3) leads directly to the Lippmann equation:

$$\left(\frac{\partial\gamma}{\partial E}\right)_{\mu_{salt}, P, T} = -\sigma_0 \tag{7.5}$$

which was put forward over a century ago and is the basis of the study of **electrocapillarity**. The interfacial tension can be measured in a variety of ways (§6.3.2) of which the most common for the dropping mercury electrode (DME) is the drop weight method. The high surface tension of mercury makes the drop almost perfectly spherical in shape and by weighing a known number of drops the (relative) value of γ can be estimated.

Just before the drop falls from the capillary, its weight is balanced by the surface tension force (Fig. 6.12(a)):

$$2\pi r\gamma = Mg \tag{7.6}$$

where r is the (external) radius of the capillary. Although this equation is only approximate, it remains true that the value of γ is *proportional* to the

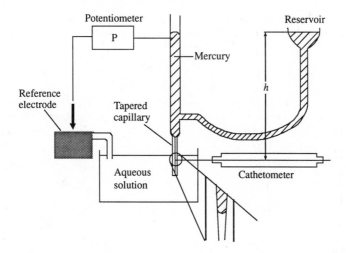

Fig. 7.4. The Lippmann electrometer. The capillary tube at the base of the left-hand mercury column is slightly tapered.

drop mass under different conditions (with the same capillary) so values of γ can be obtained by comparison with the known value in, say 0.01 M KCl, with some specified applied potential.

Alternatively, in the capillary electrometer the mercury–electrolyte interface is formed in a tube of varying radius (Fig. 7.4). As γ varies with E, the hydrostatic pressure (measured by h) required to return the interface to the same position in the capillary is measured. Since that is determined by the diameter at that point (and γ) one can calculate γ (or a relative γ) from the Young–Laplace equation (5.6).

Note that when there is no charge on the mercury surface $(\partial\gamma/\partial E)_\mu = 0$ and at this value of E the interfacial tension is a maximum. Figure 7.5 shows some typical plots of γ against the applied potential difference. When E is made negative (**cathodic polarization**), the mercury surface is negative with respect to the solution. The predominant counterions on the solution side are then the cations and Fig. 7.5 shows that Na^+ and K^+ behave essentially identically with respect to γ. On the other hand, under **anodic polarization**, the counterions are anions and it is clear from the figure that even halide ions behave very differently from one another. It is hardly surprising that anions should show some of their *chemical* character against a metal surface whereas cations tend to behave more like positive charges, simply responding to the local electrostatic potential. We will return to this point later.

The most striking feature of the curves in Fig. 7.5 is the maximum in γ, called the **electrocapillary maximum** (e.c.m.). The maximum identifies the

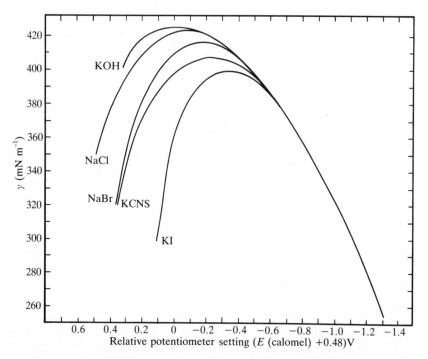

Fig. 7.5. Electrocapillary curves (reprinted with permission from Grahame (1947)). The potentiometer reading E has been adjusted so that $E = 0$ at the e.c.m. for sodium fluoride. (Copyright American Chemical Society.)

point where σ_0 is zero from eqn (7.5) and so it is also referred to as the **point of zero charge** (p.z.c.). Putting a charge on the surface, whether positive or negative, has the same effect as adsorbing a surfactant at other liquid surfaces (§6.3.3).

The charge at any other value of E can be obtained by differentiation of the γ versus E curve (eqn (7.5)), but to get accurate values requires very accurate initial data. At the dropping mercury electrode an alternative method of proceeding is to determine the **electrical capacitance** of the mercury surface. The capacitance measures the amount of charge which can be stored on the mercury surface (per unit area) for a given potential difference across the interface. (Generally speaking, the bigger the applied potential, E, the bigger the stored charge, but some systems can store a lot more charge than others for a given potential difference; it is in that sense that they have a higher capacity.)

We noted earlier that the arrangement of a charge on the metal and an opposite charge in the solution behaves like a parallel plate capacitor. By surrounding the drop with a counterelectrode, in the form of a cylinder of

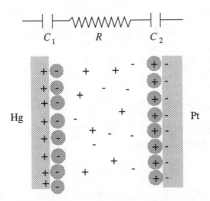

Fig. 7.6. Electrical equivalent circuit for the dropping mercury electrode.

platinum gauze (Fig. 7.1) one can determine the total impedance to the flow of an alternating current between the platinum and the mercury. That impedance is made up of a capacitor on each of the electrodes with a resistor in between (corresponding to the conduction of the current through the electrolyte) (Fig. 7.6). The system behaves electrically as two capacitors in series with a resistor.

The impedance of a capacitor is inversely proportional to its capacitance, and since the area of the platinum gauze is much greater than the area of the mercury drop the capacitance C_2 is much greater than C_1 and so its impedance is very much smaller. It can usually be neglected so that the system behaves like a resistor, R, in series with a capacitor (C_1). A good conductance-measuring bridge can easily determine the best pair of values for R and C_1.

There remains the problem caused by the fact that as the drop of mercury grows in size, the capacitance increases and so it is necessary to find the exact time in the life of the drop at which the bridge is balanced. That can be done by setting the bridge to values for C and R which are known to occur some time in the life of the drop and then following the out-of-balance signal and determining the exact time when it passes through zero (that is, bridge balance). A drop normally takes about 10 s to grow and drop off and it is relatively easy to determine the time between birth and balance to within about 1 ms and to estimate the area of the drop at that instant. Thus one can estimate the effective capacitance per unit area of the mercury surface.

The measured value is actually the **differential capacitance** which is defined as:

$$C = \left(\frac{\partial \sigma_0}{\partial E}\right)_\mu \tag{7.7}$$

and which, from eqn (7.5), must be equal to $-d^2\gamma/dE^2$. The importance of the capacitance is that, being a differential quantity, it contains more reliable information than does the surface tension.

To see this more clearly we note that the curves in Fig. 7.5 are very nearly parabolic. If they were represented by an equation of the form

$$\gamma = \gamma_{ecm} - b(E - E_{ecm})^2 \tag{7.8}$$

(where the subscript ecm refers to the electrocapillary maximum) we would have (from eqn (7.5)):

$$-(d\gamma/dE) = 2b(E - E_{ecm}) = \sigma_0 \tag{7.9}$$

and then $C = d\sigma_0/dE = 2b$ (that is, a constant). The actual experimental values are far from constant, as is shown in Fig. 7.7. These data are for sodium fluoride, which is now recognized as the simplest possible electrolyte behaviour to interpret.

The data in Fig. 7.7 can be used to determine the charge on the electrode at any value of E since (from eqn (7.7))

$$\sigma_0 = \int_{E=E_{ecm}}^{E} C\, dE. \tag{7.10}$$

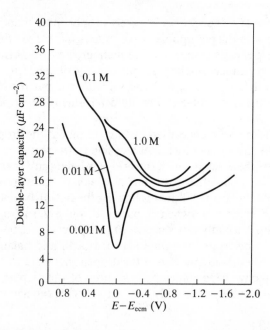

Fig. 7.7. Differential capacitance at the dropping mercury electrode in contact with NaF solutions at 25 °C (reprinted with permission from Grahame (1947)). (Copyright American Chemical Society.)

A second integration can be used to evaluate γ from the Lippmann equation (7.5):

$$\gamma - \gamma_{\text{ecm}} = -\int \sigma_0 \, \mathrm{d}E = -\int_{E_{\text{ecm}}}^{E} \int C(\mathrm{d}E)^2 \qquad (7.11)$$

and this value can be compared with the direct measurement as a check on the reliability of the data.

If measurements are done at a variety of salt concentrations (as shown in Fig. 7.7), one can determine the values of Γ_- from eqn (7.3):

$$\Gamma_- = -(\partial\gamma/\partial\mu)_E \qquad (7.12)$$

and hence Γ_+ from eqn (7.4). In this way a large body of data has been built up on the relation between the electrical potential on the mercury surface and the interfacial adsorption for a large number of ionic systems.

We have looked at the mercury surface in some detail because the ideas which have been developed there have been carried over to the solid surfaces more commonly encountered in colloid science. On those surfaces, the notion of a surface tension is not of much help, but the equilibrium surface energy can be inferred from the measured surface charge using the fundamental thermodynamic equations (for example eqn (7.11)); the surface charge density is quite easy to measure in a solid/liquid system.

7.2 Charge and potential distribution near the mercury surface

The curves in Fig. 7.7 have two important general characteristics. At low salt concentrations there is a minimum in the capacitance at the e.c.m. or point of zero charge. At high concentrations, the capacitance falls more or less regularly as the potential falls from positive values through zero to negative values; that is as the counterions change from being anions to cations. The picture we have of the mercury solution interface is based on those two facts. The capacitance of the interface (as shown in Fig. 7.7) is thought to consist of two capacitors in series: one dominates the behaviour at low salt concentration and low potentials, the other is important at high salt concentration or when the potential is high.

The shape of the two is shown in Fig. 7.8. Remembering that when two capacitors are in series it is the smaller one which is most important, we can see why the shape of the curves changes as shown in Fig. 7.7. Taking always the value of the lowest capacitance we get the full line shapes shown in Fig. 7.8 at low and high salt concentration respectively and they reproduce the general features of the capacitance curves of Fig. 7.7.

To understand how these two capacitors arise and why they have the shapes they do, we need to analyse the relation between electrostatic potential and charge in the neighbourhood of the interface. That requires some

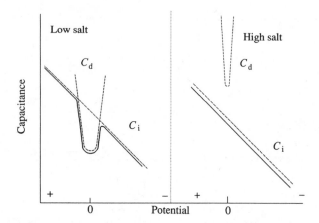

Fig. 7.8. The two series capacitors, C_i, and C_d, which determine the total capacitance at the mercury–solution interface. The total capacitance expected with this model is shown by the full line.

basic physics and the solution of one of the most important equations in all of physical chemistry: an equation which is used for systems as diverse as electrolyte solutions and high-temperature plasmas, protein solutions and the properties of porous media and membranes, as well as the theory of transistors, electrode processes, and colloid stability.

The equation we seek will link the electrostatic potential at each point near a metal surface to the amount of electrical charge in the neighbourhood. The charge on the metal is assumed to be spread evenly over its surface; the metal surface is a plane of constant potential. In the solution, the counterions cannot quite reach up to the metal surface because of their finite size. There is, therefore, a charge-free region right near the metal surface and this forms the capacitor C_i. Outside that region, the **diffuse double layer** begins and it is in this diffuse layer that the equation is assumed to apply. The charge will be in the form of ions but we will have to assume that those ions are point charges of magnitude ze which respond to the local electrostatic potential.

7.2.1 The Poisson–Boltzmann equation

We showed in Fig. 2.8 what the distribution of potential looks like in the neighbourhood of a colloidal particle. We now want to show how that potential distribution arises.

The earliest studies of the electrical double layer near a metal surface, by Helmholtz in the mid-nineteenth century, assumed that the electric charge on the metal surface was balanced by an equal charge which sat in the surrounding solution a little way from the surface. When the kinetic molecular theory of matter became established, towards the end of the nineteenth century, it was obvious that that model was seriously defective. The charges in the solution were on the ions and they would be moving about as a result of their thermal energy. Around 1910 a better model was proposed by a French scientist named Gouy and a similar treatment was developed independently a few years later by the British scientist, Chapman. The result is known as the **Gouy–Chapman** model of the electrical double layer. It assumes that the electrical charge on the metal surface influences the distribution of ions in the electrolyte, so that an excess of ions of opposite sign is established in the layers of solution near the surface.

The electrostatic potential, ψ, at any point near the surface is related to the net number of electrical charges per unit volume in the neighbourhood of that point; the **volume charge density**, ρ, measures the excess of positive over negative ions, or vice versa. Both ψ and ρ are quantities which are averaged over a sufficiently long time to eliminate fluctuations due to the motion of the ions. (That will be a time of order microseconds in most cases.)

According to the theory of electrostatics, the relation between ψ and ρ is given by the Poisson equation, which, for a flat surface, can be written:

$$d^2\psi/dx^2 = -\rho/\varepsilon_w = -\rho/\varepsilon_0\varepsilon_r \qquad (7.13)$$

where ε_w is the permittivity of water which can be written in terms of the relative permittivity (or dielectric constant) ε_r and the permittivity of a vacuum, ε_0. This assumes that the potential in planes parallel to the surface is constant and only the distance, x, at right angles to the metal surface is involved in determining the potential (Fig. 7.9).

The ions are influenced by the local electrostatic potential as well as by

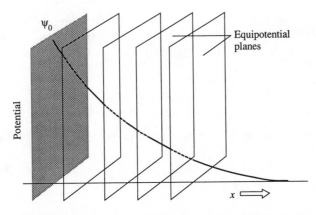

Fig. 7.9. Electrostatic potential is constant in planes parallel to the metal surface.

their thermal energy. The number of ions of each type, n_i, is then given by the Boltzmann equation (compare eqn (2.5)):

$$n_i = n_i^0 \exp\left(-w_i/kT\right) \qquad (7.14)$$

where n_i^0 is the number of ions of type i per unit volume in the bulk solution far from the surface, k is the Boltzmann constant, and T is the temperature (in kelvin). The work done (or the extra energy involved) in getting an ion to the point in question, w_i, is assumed to be measured simply by the electrostatic energy it acquires: $w_i = z_i e\psi$, where ψ is taken to be zero in the bulk solution far from the surface. Thus we see that ions with the same sign as the surface charge will be repelled from the interface ($n_i < n_i^0$) and ions of opposite sign will be attracted. The volume density of charge, ρ, is obtained by adding up all the ions of either sign in a unit volume of the electrolyte solution in the neighbourhood of the point in question:

$$\rho = \sum_i n_i z_i e = \sum_i n_i^0 z_i e \exp\left(-z_i e\psi/kT\right). \qquad (7.15)$$

The Poisson–Boltzmann equation for this particular situation can then be written:

$$\frac{d^2\psi}{dx^2} = \frac{-1}{\varepsilon_0 \varepsilon_r} \sum_i n_i^0 z_i e \exp\left(-z_i e\psi/kT\right). \qquad (7.16)$$

This is a second-order nonlinear differential equation which can be solved exactly, but before doing so we will examine a useful approximation.

7.2.2 The Debye–Hückel approximation

If the electrical energy is small compared with the thermal energy of the ions ($|z_i e\psi| < kT$) it is possible to expand the exponential in eqn (7.16) ($e^{\pm x} = 1 \pm x + x^2/2! + \ldots$), neglecting all but the first two terms, to give:

$$\frac{d^2\psi}{dx^2} = -\frac{1}{\varepsilon_w}\left[\sum_i z_i e n_i^0 - \sum_i z_i^2 e^2 n_i^0 \psi/kT\right]. \qquad (7.17)$$

The first summation term must be zero to preserve electroneutrality in the bulk solution, so

$$\frac{d^2\psi}{dx^2} = \left(\frac{\sum z_i^2 e^2 n_i^0}{\varepsilon_w kT}\right)\psi = \kappa^2\psi \qquad (7.18)$$

where

$$\kappa = \left(\frac{e^2 \Sigma\, n_i^0 z_i^2}{\varepsilon kT}\right)^{1/2}. \tag{7.19}$$

This simplification of assuming ψ to be very small is called the Debye–Hückel (linear) approximation because it was used by those two scientists in their very successful theory of strong electrolytes. Equation (7.18) can readily be shown (Exercise 7.2.1) to have a solution of the form

$$\psi = \text{constant} \times \exp(-\kappa x). \tag{7.20}$$

The quantity κ (which has the dimensions (length)$^{-1}$) is called the **Debye–Hückel parameter** and it plays an important part in the theory of the double layer. The quantity $1/\kappa$ is often referred to as the '**thickness of the double layer**'. In fact, the region of varying potential shown in Fig. 7.9 is of the order of $3/\kappa$ to $4/\kappa$. Note that, apart from fundamental constants, κ depends only on the temperature and the bulk electrolyte concentration. At 25°C in water the value of κ is given by (Exercise 7.2.1):

or
$$\kappa^2 = (2000 F^2/\varepsilon_0 \varepsilon_r RT) I \tag{7.21}$$

$$\kappa = 3.288\sqrt{I}\,(\text{nm}^{-1})$$

where F is Faraday's constant. I is the **ionic strength** of the solution:

$$I = \tfrac{1}{2}\sum (c_i z_i^2)$$

where c_i is the solution concentration (mol l^{-1}). In 10^{-3} M 1:1 aqueous electrolyte solution, $1/\kappa$ is 9.6 nm and for systems of interest in colloid science, $1/\kappa$ ranges from a fraction of a nanometre to about 100 nm (Exercise 7.2.2). As the ionic concentration increases, the 'thickness' of the double layer decreases, a process referred to as **compression of the double layer**.

7.2.3 Solution of the complete Poisson–Boltzmann equation

Unfortunately, in many situations of interest in colloid science and electrochemistry†, the assumption that $|ze\psi| < kT$, does not hold. The range of values of E shown in Figs. 7.5 and 7.7 suggests that we are dealing with electrostatic potentials of the order of 1 V, so that $e\psi \approx 1.6 \times 10^{-19}$ J, which is about $40kT$ at room temperature. Under such conditions the complete Poisson–Boltzmann equation must be solved. Fortunately, that is not too difficult for a flat surface, though it does involve some manipulation of hyperbolic functions.

---

† Except in the theory of strong electrolytes. The potential falls off very quickly near an ion and because the ions are of finite size they cannot get too close to one another. That is why the Debye–Hückel approach works reasonably well there.

To simplify the algebra we set $z_i = z_+ = -z_- = z$ so that the analysis is limited to symmetrical $z{:}z$ valent electrolytes. It turns out that this is not a very serious restriction because in most situations of interest in colloid science the behaviour is determined overwhelmingly by the ions of sign opposite to that of the surface (see §2.3.5). Equation (7.16) can then be written (Exercise 7.2.3):

$$\frac{d^2\psi}{dx^2} = \frac{2n^0 ze}{\varepsilon} \sinh \frac{ze\psi}{kT} \tag{7.22}$$

using the identity $\sinh p = [\exp p - \exp(-p)]/2$.

This can be integrated by multiplying both sides by $2(d\psi/dx)$:

$$\frac{2d\psi}{dx}\frac{d^2\psi}{dx^2} = \frac{4n^0 ze}{\varepsilon} \sinh \frac{ze\psi}{kT} \frac{d\psi}{dx}. \tag{7.23}$$

The left-hand side is the differential (with respect to x) of $(d\psi/dx)^2$. Integrating with respect to x then gives

$$\int \frac{d}{dx}\left(\frac{d\psi}{dx}\right)^2 dx = \int \frac{4n^0 ze}{\varepsilon} \sinh \frac{ze\psi}{kT} \, d\psi. \tag{7.24}$$

Integrating from some point in the bulk solution where $\psi = 0$ and $d\psi/dx = 0$ (Fig. 7.9) up to a point in the diffuse double layer we obtain

$$\left(\frac{d\psi}{dx}\right)^2 = \frac{4n^0 kT}{\varepsilon}\left(\cosh \frac{ze\psi}{kT} - 1\right) \tag{7.25}$$

or (Exercise 7.2.4)

$$\frac{d\psi}{dx} = -\left(\frac{8n^0 kT}{\varepsilon}\right)^{1/2} \sinh \frac{ze\psi}{2kT}$$

$$= -\frac{2\kappa kT}{ze} \sinh \frac{ze\psi}{2kT} \tag{7.26}$$

using eqn (7.19). Note that the negative sign is chosen so that $d\psi/dx$ is always negative for $\psi > 0$ and positive for $\psi < 0$. This ensures that $|\psi|$ always decreases going towards the bulk solution and becomes zero far from the surface.

We now integrate eqn (7.26) from the bulk solution up to a point a little distance, d, from the metal surface. (The region from the surface out to d is the charge-free region we referred to earlier, and the diffuse double layer is assumed to start at a distance d from the surface.) If the potential at that point is ψ_d, the result is (Exercise 7.2.5):

$$\tanh(ze\psi/4kT) = \tanh(ze\psi_d/4kT)\exp[-\kappa(x-d)]. \tag{7.27}$$

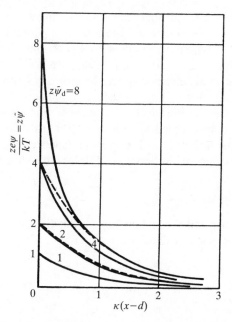

Fig. 7.10. Electrostatic potential in the diffuse double layer according to the Gouy–Chapman model. Full curves are from eqn (7.27) whilst broken curves are from eqn (7.28) for $ze\psi_d/kT = 2$ and 4. ($\tilde{\psi}(= e\psi/kT)$ is a dimensionless quantity called the **reduced potential**.)

For very low potentials the substitution $\tanh p \approx p$ can be made and eqn (7.27) reduces to:

$$\psi = \psi_d \exp\left[-\kappa(x - d)\right] \tag{7.28}$$

which is the solution of the linear equation (7.18). This is often referred to as the Debye–Hückel approximation.

A comparison between eqns (7.27) and (7.28) is shown in Fig. 7.10. Notice that for $(ze\psi/kT) < 2$ (that is $|z\psi_d| < 51.4\,\text{mV}$ at room temperature) the approximation is reasonably accurate. Also, at relatively large distances from the surface, we can substitute $\tanh p \approx p$ and then

$$\psi = (4kT/ze)\, Z \exp\left[-\kappa(x - d)\right] \tag{7.29}$$

where

$$Z = \tanh \frac{ze\psi_d}{4kT}$$

$$= \frac{\exp\left(ze\psi_d/2kT\right) - 1}{\exp\left(ze\psi_d/2kT\right) + 1}. \tag{7.30}$$

Since Z approaches unity for high values of $z\psi_d$ one can expect the potential far from the wall to resemble that from a wall of potential $\psi_d = 4kT/ze = (4 \times 25.7/z)$ mV, *irrespective* of the actual potential, provided it is sufficiently high (compare eqn (7.29) with (7.28)). In colloidal situations, any measurement on a highly charged system at ordinary temperatures will suggest that $\psi_d \approx (100/z)$ mV if the measurement method only samples the outer regions of the double layer. Likewise, if one wants to predict the behaviour of a highly charged system in a situation where only the outer region of the diffuse double layer is important (such as in the flow behaviour of a dilute sol), the approximation $\psi_d \approx (100/z)$ mV should be a good one.

7.2.4 The diffuse layer charge

The total charge, per unit area of surface, in the diffuse layer is obtained by summing the volume charge density through the whole region from distance d to ∞:

$$\sigma_d = \int_d^\infty \rho\, dx \tag{7.31}$$

and substituting for ρ from eqn (7.13):

$$\sigma_d = \int_\infty^d \varepsilon \frac{d^2\psi}{dx^2}\, dx = \varepsilon \left(\frac{d\psi}{dx}\right)_\infty^d. \tag{7.32}$$

Now $(d\psi/dx)_{x\to\infty} = 0$, so that $\sigma_d = \varepsilon\,(d\psi/dx)_{x=d}$ and using eqn (7.26):

$$\sigma_d = \frac{-2\kappa kT\varepsilon}{ze} \sinh \frac{ze\psi_d}{2kT}$$
$$= \frac{-4n^0 ze}{\kappa} \sinh \frac{ze\psi_d}{2kT}. \tag{7.33}$$

Note that the sign of σ_d is opposite to that of ψ_d (since $z > 0$).

For a symmetric electrolyte in water at 25°C, eqn (7.33) gives (Exercise 7.2.6)

$$\sigma_d = -0.1174\, c^{1/2} \sinh\,(19.46\, z\psi_d) \tag{7.34}$$

in $C\,m^{-2}$ when ψ_d is in V and c is in mol l^{-1}.

For very small potentials, where the linear equation (7.28) can be used, a similar analysis leads to (Exercise 7.2.7)

$$\sigma_d = -\varepsilon\kappa\psi_d \tag{7.35}$$

so that

$$-\sigma_d/\psi_d = K_d = \varepsilon\kappa. \tag{7.36}$$

The quantity K_d is called the **integral capacitance** of the (diffuse) double layer. Equation (7.36) shows that, for low potentials, the diffuse layer behaves like a parallel plate capacitor with a spacing of $1/\kappa$ between the plates, a charge of $+\sigma_d$ and $-\sigma_d$ on them and a potential difference of ψ_d. For low potentials then, the Helmholtz model, with the metal charge balanced by a layer of charges of opposite sign in the solution, is quite satisfactory for many purposes, even though the charge is by no means stationary in the solution. (Note that, for low potentials, the integral capacitance is the same as the differential capacitance $(-d\sigma_d/d\psi_d)$.)

A further important quantity is the differential capacitance of the diffuse layer, C_d, defined by (from (7.33)):

$$C_d = \frac{-d\sigma_d}{d\psi_d} = \frac{2n^0 z^2 e^2}{\kappa kT} \cosh\left(ze\psi_d/2kT\right)$$

$$= \varepsilon\kappa \cosh\left(ze\psi_d/2kT\right) \tag{7.37}$$

$$= 2.285\, zc^{1/2} \cosh\left(19.46\, z\psi_d\right) \mathrm{F\,m}^{-2}$$

in water at 25°C, for c in mol l^{-1} and ψ_d in V. Values of C_d, even at modest electrolyte concentrations, rapidly become very large as the potential increases from its value of zero at the p.z.c (§7.1). This is the origin of the very steep curves labelled C_d in Fig. 7.8; these may be calculated using eqn (7.37).

. It is obvious from these curves that the linear theory, giving rise to a constant capacitance, breaks down very rapidly as the electrical potential increases. That is why it is essential to introduce the complete equation to deal with most electrochemical and colloidal situations. The great advantage is that with an analytical solution like eqn (7.27) (even if it is a bit messy) we can calculate how good the linear approximation is. It often turns out in colloid science that the experimental behaviour is well described by the linear theory, even in regions where it ought not to hold. This must be a consequence of the fortuitous cancellation of effects which tend to push the results in opposite directions.

7.2.5 The inner (compact) double layer

The region between the metal surface and the beginning of the diffuse double layer was described as free of ions and it was necessarily there because all ions have a finite size, so their centres can only approach to within a distance of one ion radius. When we consider the problem in more detail we see that the radius displayed by the ion will depend on whether or not it remains hydrated when it approaches the interface. It turns out that, at the metal-solution interface, cations tend to remain hydrated whereas many anions do not. The result is that the region which denies access to the cations may contain some

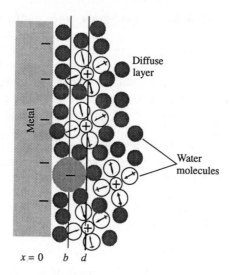

Fig. 7.11. The double-layer structure at the mercury–solution interface according to the usually accepted model.

anions, depending on whether or not they are strongly attracted to the metal surface (Fig. 7.11). It is referred to as the inner or 'compact' part of the double layer to distinguish it from the outer 'diffuse' layer; compact and diffuse refer to the arrangement of the ions in the layer. The compact layer is normally about 0.5 nm thick.

If there is no charge between the metal and the beginning of the diffuse double layer, then electroneutrality requires that

$$\sigma_0 = -\sigma_d. \tag{7.38}$$

This inner double-layer region behaves as a parallel plate condenser or capacitor with a charge of σ_0 on one plate and σ_d on the other and a potential drop across the gap of $\psi_0 - \psi_d$. If the thickness of this layer is d then the capacitance is given by

$$C_i = K_i = \sigma_0/(\psi_0 - \psi_d) = \varepsilon_i/d. \tag{7.39}$$

where the permittivity of this inner region, ε_i, is, in general, not the same as that of normal water. Indeed it may vary somewhat across the layer, and then ε_i has to be a suitable average value but it normally has a relative value of 6–20 compared with the normal value of nearly 80 for ordinary water. This is because the water molecules in this region are strongly oriented and are not able to react to an applied electric field in the same way as 'normal' bulk water. The result is that the capacitance of the inner layer is much less where ψ_i is the potential in the plane $x = b$ where the adsorbed ions are assumed to be placed and ε_1 and ε_2 are the (average) permittivities on the left and the right of that plane respectively. Charge balance now requires that

$$\sigma_0 + \sigma_i + \sigma_d = 0. \tag{7.42}$$

It is common practice to assume that ε_1 and ε_2 are constant, so that eqns (7.40) and (7.41) predict a linear change in potential in each region (Fig. 7.12), but that is not a necessary restriction. One can take a more realistic view with these permittivities varying smoothly in the compact region, and then the potential too changes smoothly in going from the metal surface to the beginning of the diffuse layer.

7.2.6 *Mathematical description of the charged interface*

A complete solution of the potential and charge distribution in the double layer would provide us with values of the six variables: the values of σ and ψ at the three planes ($x = 0$, b, and d). So far we have only four equations linking these quantities together: (eqns (7.33), (7.40), (7.41), and (7.42)). The equations also contain a number of adjustable parameters: the distances b and d and the permittivities ε_1 and ε_2. We need two more relations to complete the scheme. The first is a relation between ψ_i and σ_i which is, in effect, an adsorption isotherm (§6.1.2), since it is concerned with the number of ions which are adsorbed into the inner layer in response to the electrostatic and chemical forces which operate on an ion in that layer. The second is a method for estimating ψ_0. That is done quite differently for the metal surface and the solid surfaces in which we are most interested, so we will save that for section 7.3.

The isotherm relation is of the form

$$\sigma_i = z_i e n_i^s = z_i e f (N_s, a_i, \psi_i, \theta_i) \tag{7.43}$$

where n_i^s is the number of ions of type i which are adsorbed in the plane $x = b$ per unit area. This is expected to be a function of: (i) the number of adsorption sites (N_s); (ii) the activity (or concentration) of the ion i, a_i (or c_i); (iii) the local electrostatic potential in the plane; and (iv) θ_i, an extra free energy term which takes account of all other special effects that the ion experiences when it is in the plane $x = b$. This will involve its special interaction with the surface atoms, which will include purely physical interactions but may also involve 'chemical' effects, that is, effects which can only be properly described by molecular orbital overlap and (partial) bond formation. The expression for n_i^s is of the form of the Boltzmann equation and can be derived from a simplified form of the Langmuir isotherm (§6.1.2):

$$\sigma_i = z_i e \, n_i^s = z_i e K x_i N_s \tag{7.44}$$

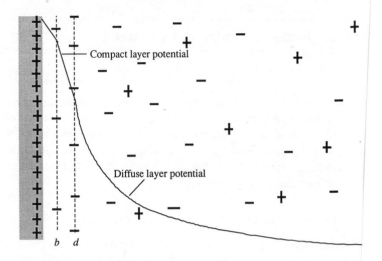

Fig. 7.12. The electrostatic potential in the compact and diffuse part: double layer.

than that of the diffuse double layer, except when the electrical pote the metal is very low (near to the e.c.m. (Fig. 7.5)). This is the pictur in Fig. 7.8.

The inner layer capacity is also affected by the value of d. W counterions are (unhydrated) anions they are able to get closer to t surface and so the value of C_i is larger than when the counter (hydrated) cations. Hence the value of C_i falls as the metal pol: goes from anodic (on the left) to cathodic (on the right)† as is s Fig. 7.8. Note that the metal surface itself is assumed to be hydrate at a point where an anion is adsorbed. This is why the assumed ca values are so much larger when anions are the counterions.

When there are ions adsorbed into the inner (compact) region, the model to describe the behaviour is that initially introduced by St structure of the **Stern layer**, as it is sometimes called, is as shown in Again the regions $0 < x < b$ and $b < x < d$ are free of charge an write equations for the potential drops across those miniature ca

$$\psi_0 - \psi_i = \sigma_0(b/\varepsilon_1)$$

and

$$\psi_i - \psi_d = -\sigma_d(d - b)/\varepsilon_2$$

† Note that this is opposite to the way we normally plot the variable on the absci convention is well entrenched in electrochemistry and is unlikely to be changed.

where x_i is the mole fraction of the ion in solution and K is the equilibrium constant for the adsorption.

Putting $\Delta G^0_{\text{ads}} = -RT \ln K$ gives:

$$\sigma_i = z_i e x_i N_s \exp\left(-\Delta G^0_{\text{ads}}/RT\right). \qquad (7.45)$$

The free energy of adsorption, ΔG^0_{ads}, in this case can be calculated assuming that the ion experiences an electrostatic attraction and some chemical effect:

$$\Delta G^0_{\text{ads}} = z_i e \psi_i + \theta_i \qquad (7.46)$$

so the adsorption equation becomes

$$\sigma_i = z_i e Q n_i^0 \exp\left[\left(-z_i e \psi_i + \theta_i\right)/kT\right] \qquad (7.47)$$

where Q is a parameter which measures the ratio of the number of ion sites on the surface to those in the bulk.

We now have six unknowns and five equations to describe the charge and potential distribution near a charged interface. If there are no ions in the Stern layer there are only four unknowns (σ_0, σ_d, ψ_0, and ψ_d) and three equations (7.33), (7.38), and (7.39). Either way we need the remaining expression for the surface potential, ψ_0.

Exercises

7.2.1 Verify that the general solution of eqn (7.18) is of the form:

$$\psi = A \exp\left(\kappa x\right) + B \exp\left(-\kappa x\right)$$

What assumptions must be made about the behaviour of ψ far from the surface for this to be a satisfactory solution? Verify the expressions in eqn (7.21) for κ using the usual values for the fundamental constants. (*Hint*: the permittivity of a vacuum, $\varepsilon_0 = 8.854 \times 10^{-12}\,\text{C V}^{-1}\text{m}^{-1}$ (or F m^{-1}).)

7.2.2 Calculate the ionic strengths and corresponding values of κ^{-1} for the following solutions: 0.01M Na_2SO_4; 0.015M $LaCl_3$; 3×10^{-3} M $Ca(NO_3)_2$; 0.025M $Fe(NH_4)(SO_4)_2$; A mixture of 10^{-4} M $La_2(SO_4)_3$ and 5×10^{-4} M $NaNO_3$. (Note that for a 1:1 electrolyte, the ionic strength is the same as the concentration.)

7.2.3 Establish eqns (7.22) and (7.25).

7.2.4 Establish eqn (7.26) using the identity:

$$\cosh p = 2\sinh^2\left(p/2\right) + 1.$$

7.2.5 Establish eqn (7.27) using the fact that: $\sinh p = 2 \sinh p/2 \cosh p/2$; $\text{sech}\, p = 1/\cosh p$; and $d \tanh p/dp = \text{sech}^2 p$.

7.2.6 Verify eqn (7.34) using the usual values of the physical constants.

7.2.7 Verify eqn (7.35) using the form of $d\psi/dx$ derived from eqn (7.28).

7.3 The surface potential of a charged solid

We discussed briefly, in §2.3.1, the various mechanisms whereby a solid surface may acquire an electric charge. One of the most important processes is the adsorption of **potential-determining ions** (p.d.i.). We noted in §2.3.1 that the classic colloidal suspension, on which so much fundamental study has been done, is the silver iodide sol, for which the p.d.i are the Ag^+ and the I^- ions. We must now examine in more detail the relation between the concentration of the p.d.i. and the resulting surface potential.

7.3.1 Potential-determining ions on silver iodide

When silver iodide crystals are placed in water, a certain amount of dissolution occurs to establish the equilibrium $AgI \rightleftharpoons Ag^+ + I^-$ and then we can write: $\mu_{AgI} = \mu_{Ag^+} + \mu_{I^-}$. The concentrations of Ag^+ and I^- in solution are very small because the solubility product is in this case very small ($K_{sp} = a_{Ag^+} a_{I^-} = 10^{-16} \approx [Ag^+][I^-]$). Nevertheless these small concentrations are very important because slight shifts in the balance between the Ag^+ ions and the I^- ions can cause a dramatic change in the charge on the surface of the AgI crystals. The surface of the crystal can be regarded as a more or less regular array of silver and iodide ions in cubic close pack and if there are exactly equal numbers of Ag^+ and I^- ions on the surface then it will be 'uncharged'. This is the p.z.c. for the silver iodide crystal. It turns out that this does *not* correspond to the point where the solution contains equal concentrations of Ag^+ and I^- ions (that is $[Ag^+] = [I^-] = 10^{-8}$). Rather, it seems that the iodide ions have a higher affinity for the surface and tend to be preferentially adsorbed. In order to reduce the charge to zero it is found that the silver ion concentration must be increased to about $10^{-5.5}$ (so that $[I^-] \approx 10^{-10.5}$ — a ratio of about 100 000 to 1). This can be done by adding a very small amount of silver nitrate solution. The charge on the crystal surface can thus be altered from highly positive, through zero, to highly negative values by the addition of small amounts (c. 10^{-6} mol l^{-1}) of silver ions (as, say, $AgNO_3$) or iodide ions (as, say, KI). The charge which is carried by the surface determines its electrostatic potential. For this reason they are called the **potential-determining ions** for the silver iodide surface. Similarly, the potential-determining ions for the barium sulphate crystal are barium and sulphate ions.

We must now calculate the **surface potential** on the crystals of silver iodide by considering the equilibrium between the surface of the charged crystal and the ions in the surrounding solution, using the **electrochemical potential**, $\bar{\mu}_i$, of the ions, as we noted in §7.1. Equilibrium requires that, for example, $\bar{\mu}$ for the silver ion be the same on the surface as it is in the bulk solution:

$$\mu_{Ag^+}(aq) + e\psi(aq) = \mu_{Ag^+}(surf) + e\psi(surf). \qquad (7.48)$$

As before we take the electrostatic potential in the bulk solution ($\psi(aq)$) to be zero or, what amounts to the same thing, set the surface potential, $\psi_0 = \psi(surf) - \psi(aq)$, and then

$$e\psi_0 = \mu^0(aq) + kT\ln a[Ag^+(aq)] - \mu^0(surf) - kT\ln a[Ag^+(surf)]. \qquad (7.49)$$

A similar equation can be written at the p.z.c., but in that case ψ_0 can be taken as zero.† Using a prime to indicate measurements made at the p.z.c.:

$$\mu^0(aq) + kT\ln a'[Ag^+(aq)] = \mu^0(surf) + kT\ln a'[Ag^+(surf)]. \qquad (7.50)$$

We now make the very important assumption that, as the surface becomes charged up by the adsorption of Ag^+ ions, *the activity of the surface silver ions does not change* (that is, $a = a'$ for the $[Ag^+(surf)]$ ions). Then subtracting eqn (7.50) from (7.49) gives:

$$\psi_0 = (kT/e)\ln(a[Ag^+(aq)]/a'[Ag^+(aq)])$$
$$\approx (kT/e)\ln([Ag^+]/[Ag^+]_{pzc}). \qquad (7.51)$$

In water at 25°C this reduces to:

$$\psi_0(volt) = 0.0598\log_{10}[Ag^+]/[Ag^+]_{pzc}$$
$$= 0.0598[pAg(pzc) - pAg] \qquad (7.52)$$

where $pAg = -\log_{10}[Ag^+]$.

Why is it reasonable to assume that $a = a'$ for the surface silver ions, that is, that the chemical activity of the silver ions on the surface is independent of the state of charge? Recall that the surface of the silver iodide crystal, even when it is 'uncharged' is covered with an array of silver and iodide ions. The silver ions are only a fraction of a nanometre apart. The few extra silver ions which are adsorbed to create the surface charge are normally negligible in number compared to the ones already there in the crystal lattice (Exercise 7.3.1). That is why we can assume that the activity will not be affected very much by their adsorption.

Equation (7.51) is called the **Nernst equation** and it shows that the surface potential increases by about 60 mV for every tenfold increase in the concentration of the silver ion. A tenfold change in the iodide concentration produces a 60 mV change in the opposite direction (Fig. 7.13).

---

† This amounts to assuming that the only important potential is that due to the adsorbed ions, which is a reasonable starting point.

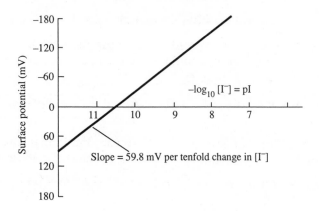

Fig. 7.13. Dependence of surface potential on iodide ion concentration.

The ease with which we can control the surface potential on silver iodide particles has made that material a very useful one for the study of the fundamental properties of colloidal systems. Unfortunately, there are many systems for which this simple procedure will not work. The crucial assumption, that the activity of the potential-determining ions on the surface is independent of the state of charge, is not always acceptable. On many oxide surfaces one does not have a lot of positive and an equal number of negative charges present at the p.z.c. The number of charged sites present at the p.z.c. is not very large at all since most of the surface–OH groups are undissociated. As the surface is charged up, by adsorption of H^+ or OH^-, the environment in which the ions find themselves is significantly changing. The Nernst equation no longer holds and the calculation of the surface potential becomes quite a difficult exercise, subject to considerable uncertainty. Before considering that situation we will look at the experimental data on the silver iodide surface.

7.3.2 *Charge, potential, and capacitance of the AgI interface*

When a silver–silver iodide electrode is dipped into a solution containing either silver or iodide ions, the potential developed at the electrode depends on the concentration of the ions in solution, according to the Nernst equation (7.51). This fact makes it possible to readily determine the amount of charge on the particles of silver iodide in a suspension, as a function of the surface potential. The Ag/AgI electrode can be used to determine the amount of Ag^+ or I^- ions in a solution, in the same way that a glass electrode is used to determine the number of H^+ ions in a solution (Fig. 7.14). The reference electrode can be a calomel half cell, just as it is for the deter-

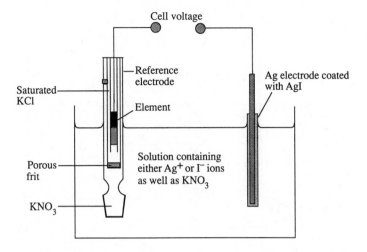

Fig. 7.14. Measurement of Ag^+ or I^- concentration with an electrochemical cell.

mination of pH. The voltage of the cell is made up of contributions from the two electrodes:

$$E_{cell} = E_{Ag/AgI} + E_{ref}.$$

Now E_{ref} is fixed and $E_{Ag/AgI}$ changes by 60 mV per tenfold change in the concentration of Ag^+ or I^- ions (eqn (7.52)). This is a simple example of an **ion-selective electrode**, of which there are many types now available. The other electrolytes present (like KNO_3 in the system shown in Fig. 7.14) do not affect the potential of the electrode and so are called **indifferent electrolytes** (§2.3.4).

Consider now a system in which a Ag/AgI electrode (and a reference electrode) is dipping into a suspension containing a few grams of colloidal silver iodide with a total surface area of, say, 20 m². If we now measure the cell voltage and then add a known amount of $AgNO_3$ to the suspension and measure the change in the voltage of the cell we can tell how many Ag^+ ions remain in the solution. Hence we can estimate how many have been adsorbed onto the surface of the solid since it is equal to the difference between the number added and the number remaining. The amount of solid AgI is chosen so that the adsorption of ions from the solution is large enough to be measured directly. Provided the surface area of the solid is known, it is an easy matter to calculate the surface density of charge (in μC per cm²). By such a **surface titration** procedure, it is possible to measure the surface charge as a function of the concentration of Ag^+ ions in solution (and hence of the surface potential using eqn (7.51)).

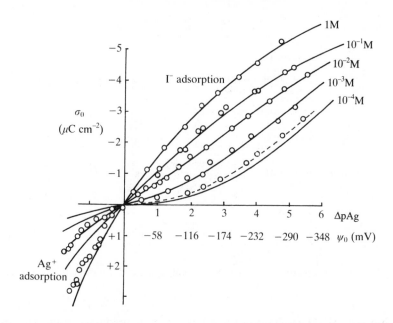

Fig. 7.15. Surface charge of silver iodide determined by titration. The concentrations shown on the graph are those of the indifferent electrolyte. $\Delta pAg = pAg - pAg_{pzc} = pAg - 5.5$. (From Overbeek (1952), with permission.)

The result of such studies is shown in Fig. 7.15. Each curve is for a different concentration of the background (indifferent) electrolyte solution, which is KNO_3 or something similar. This provides a **constant ionic strength** so that the extent of the diffuse part of the double layer (κ^{-1}) remains the same along each curve. Note that the iodide adsorption can be studied more thoroughly because the surface remains negative for a wide range of (dilute) iodide concentrations (from $[I^-] \approx 10^{-10.5}$ to $10^{-5.5}$ M). Even at the highest concentration of iodide ion it is still very dilute compared with the concentration of the background electrolyte.

On the positive side, the concentration of Ag^+ ions starts off at $10^{-5.5}$ and becomes more concentrated as the surface becomes more positive. It can only be taken up to about 10^{-4} before the concentration of the silver begins to be comparable with that of the background electrolyte. What these curves show is that in order to maintain a certain surface potential as the background (indifferent) electrolyte concentrations is increased, it is necessary to adsorb more and more of the p.d.i. onto the surface. One could say that the addition of indifferent ions into the solution drives the p.d.i. onto the surface. Equation (7.32) for the surface charge shows why this must be so. As the indifferent electrolyte concentration increases, the potential falls off

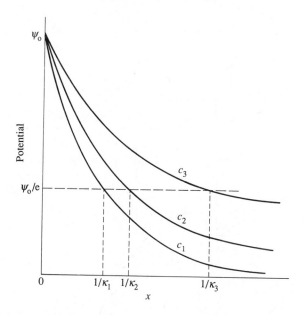

Fig. 7.16. The potential distribution near a surface at different values of the (indifferent) electrolyte concentration for a simple Gouy–Chapman model of the double layer. ($c_1 > c_2 > c_3$.) For low potential surfaces, the potential falls to ψ_o/e at distance $1/\kappa$ from the surface.

more rapidly at the surface (Fig. 7.16). The slope of that curve ($d\psi/dx$) at the surface is proportional to the surface charge.

Looking at it another way, as the salt concentration is increased, the surface charge becomes better screened by the salt ions. The work involved in bringing another charge up to the surface is therefore decreased and that work is the measure of the surface potential. To keep the potential the same, one must increase the surface charge so that the amount of work to be done in bringing a new charge up to the surface remains the same.

There is one important point which has been glossed over in the above analysis. When we measure the amount of adsorbed Ag^+ or I^- ion on the silver iodide surface, we can only determine the amount which is *added* to what is already there. To obtain the *absolute* charge we need to know when the charge is zero. The principle of the method is shown in Fig. 7.17. When a plot is made of the relative surface charge density as a function of the concentration of the potential-determining ion, at different values of the indifferent electrolyte concentration, we usually find that the curves have a common intersection point. It can be shown (Hunter 1987) that that point corresponds to the point of zero charge and so the absolute charge is obtained by adjusting all the curves as indicated in Fig. 7.17.

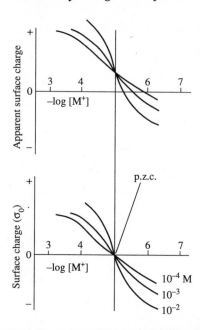

Fig. 7.17. Conversion of relative to absolute surface charge density. The common intersection point is identified as the point of zero charge and the remaining curves are adjusted accordingly.

Once the charge and potential are known, we can calculate the capacitance of the interfacial region. Then, using the same assumptions as were used for the mercury–solution interface, we can estimate the capacitance of the inner (compact) part of the double layer. That turns out to be much more charge dependent than on the mercury surface. The values vary almost linearly from a high of around $50 \ \mu F \ cm^{-2}$ down to about $10 \ \mu F \ cm^{-2}$ as the charge on the surface changes from $+2$ to $-4 \ \mu C \ cm^{-2}$. Most of that change can be accounted for by the change in the thickness of the compact layer. As the surface changes from positive to negative, the counterion changes from an anion to a cation. The distance of closest approach of the (hydrated) cation is larger than that of the anion and so the inner layer capacitance is lower (eqn (7.39)).

7.3.3 Other solid surfaces

Some other solid surfaces behave like the classical silver iodide surface, having a potential which obeys the Nernst equation (7.51). The other silver halides, silver sulphide, calcium carbonate, and calcium phosphate fall into this category. But the great majority of important solid surfaces do not. The metal oxides and hydroxides in particular do not obey the Nernst equation and that can be easily understood. The fundamental assumption involved in deriving that equation (that is that the activity of the p.d.i. on the surface remains constant as the surface is charged up (§7.3.1)) is not true for most oxides; TiO_2 is a possible exception.

As we noted in §2.3, the oxide surface becomes charged by the interaction of surface groups with hydrogen and hydroxyl ions:

$$M\text{-}OH + OH^- \rightarrow M\text{-}O^- + H_2O$$
$$M\text{-}OH + H^+ \rightarrow MOH_2{}^+. \tag{2.25}$$

At the point of zero charge on the oxide there are very few charges present. Most of the surface is covered with uncharged hydroxyl groups, whereas in AgI there are equal numbers of positive and negative ions. As the oxide surface is charged up, the charging ions experience a changing environment as the charge grows. We cannot assume that the activity is independent of the state of charge, and we should not be surprised that the Nernst equation is not obeyed. In general, the surface potential changes more slowly than the Nernst equation predicts, as we move away from the p.z.c. In other words, the surface potential changes by less than 60 mV for each unit change in pH. To estimate the surface potential as a function of the pH we need to know more about the properties of the surface groups, in particular the dissociation constant or constants.

The same remarks apply to protein-covered surfaces and, hence, to most biological surfaces. There are two distinct groups with different acid or base dissociation constants. There is also now available a wide range of ionic surface active agents which can be relied upon to adsorb strongly onto the surface of a colloidal particle and so to confer a charge on it, or to modify the charge already present. Such substances are widely used to control the charge status and, hence, the stability (in the colloidal sense (§2.3)) of the system. Adsorbed protein and natural polysaccharides (gums) have been used for centuries to stabilize colloidal dispersions, like paint pigments, inks, cosmetics, and food preparations.

These more complex adsorbates are best regarded as **specifically adsorbed** rather than as contributors to the **surface charge**. They are assumed to be located in the inner (compact) part of the double layer and are not potential-determining ions, even though they dramatically affect the potential of the diffuse part of the double layer.

To introduce the effect of surface dissociation, we will examine the simplest possible case: a surface with only one type of surface group, characterized by an acid dissociation constant K_a:

$$AH \rightleftharpoons A^- + H^+$$

with

$$K_a = \frac{[A^-]a(H^+)_s}{[AH]}. \tag{7.53}$$

The activity of protons on the surface, $a(H^+)_s$, is given by a Boltzmann equation:

$$a(H^+)_s = a(H^+)_b \exp(-e\psi_0/kT) \tag{7.54}$$

where the subscript b refers to the bulk solution. The surface charge, σ_0, is given by

$$\sigma_0 = -e[A^-]. \tag{7.55}$$

If the total number of surface sites is $N_s = [A^-] + [AH]$, then it is not difficult to show that (Exercise 7.3.2)

$$-\sigma_0 = \frac{eN_s}{1 + (a_b/K_a)\exp(-e\psi_0/kT)}. \tag{7.56}$$

Then, in the absence of specific adsorption (from eqn (7.34)):

$$\sigma_d = -\sigma_0 = -0.1174\sqrt{c}\sinh(19.46\,\psi_d) \tag{7.57}$$

where c is in mole l^{-1} and the charge is in $C\,m^{-2}$. (This is calculated for water at 25°C.) Equations (7.56) and (7.57) can be solved simultaneously to find an equation for ψ_0, provided we assume that $\psi_0 \approx \psi_d$. (That is a rather drastic assumption, but it will bring out one important point and that is how different the behaviour of this surface is from the behaviour expected from the Nernst equation.) Equation (7.56) can be written:

$$\left(\frac{a_b}{K_a}\right)\exp(-e\psi_0/kT) = -\frac{eN_s}{\sigma_0} - 1. \tag{7.58}$$

Taking logs of both sides and using $pH = -\log_{10}a(H^+)_b$ and $pK_a = -\log_{10}K_a$:

$$\psi_0 = 0.0598\left[(pK_a - pH) - \log_{10}\left(\frac{-eN_s}{0.1174\sqrt{c}\sinh(19.47\psi_0)} - 1\right)\right]. \tag{7.59}$$

Note that $d\psi_0/dpH$ would be 0.0598 V, if this system obeyed the Nernst equation. That corresponds to the first term in eqn (7.59). The actual rate of change is significantly modified by the second term in the equation, which depends strongly on the indifferent electrolyte concentration.

Exercises

7.3.1 Calculate the surface charge density on a silver iodide crystal in 10^{-3} M KNO_3 at a surface potential of 60 mV. How many silver ions are there per cm^2? Compare this with the number of Ag^+ ions in the crystal lattice (say 0.4 nm apart).

7.3.2 Establish eqn (7.56).

7.4 Modelling the solid–solution interface

A great deal of effort has gone into the task of determining the distribution of electric charge and electrostatic potential in the immediate neighbourhood of the surface of a colloid particle. The reason for this is that most of the transport properties, like electrical conductivity and diffusion coefficient and the flow properties are, in many systems, ultimately determined by that charge distribution. The difficulty is that we have direct experimental access to only certain features of the charge and potential distribution and they are not always the most important aspects. Those have to be inferred by constructing a good model of the interface. The better the model, the better we will be able to predict the behaviour of the system under different imposed conditions. We may want to estimate the effect of sea salt on the particulate material brought down by a river, or the leakage of material from the tailings dam of a gold mine, or the migration of radioactive particles from a nuclear waste dump. All these and many more such situations call for a good model of the solid/solution interface and it must be said that we still have a long way to go before we will be satisfied with our models, even for well studied systems. This is one of the most important areas of the subject and one which still largely eludes us.

Space does not permit any further elaboration of the current ideas here, but suffice it to say that for an oxide surface it is currently considered necessary to treat the surface as being characterized by groups which have two acid dissociation constants (corresponding to the acid and base amphoteric nature of the M–OH groups) and to assume as well that both the cation and the anion of the *indifferent* electrolyte are strongly bound to surface groups with a well defined dissociation constant or complexing constant. Those ions are assumed to occupy the inner (Stern) layer next to the particle surface. With all that we are still not able to match experimental behaviour exactly, as can be seen from Fig. 7.18 which represents some of the best available data in a system which has been studied much more carefully than any other. Although the charge is well described, the potential is not. The zeta potential shown in this figure is a measure of the diffuse layer potential ψ_d. It is determined by measuring the velocity of the particles in an electric field and will be described in more detail in Chapter 8. Introducing

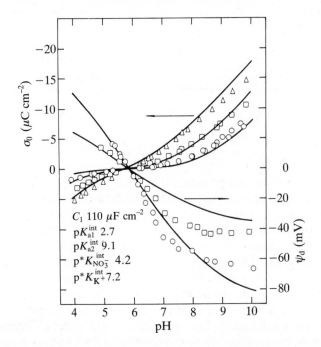

Fig. 7.18. Surface charge density and zeta potential of a TiO_2 dispersion as a function of pH at various concentrations of KNO_3 and at 25 °C. Full curves are calculated from the site binding model with constants indicated on the figure. $\triangle$, 10^{-1} M; $\square$, 10^{-2} M; $\bigcirc$, 10^{-3} M. (From James and Parks (1982).)

the zeta ($\zeta -$) potential as a measurable quantity enables us to get around some of the difficulties discussed above. The ζ-potential is, in many cases, a good estimate of the diffuse layer potential (ψ_d) and can be used to estimate the likely effect of various reagents on the properties of the colloidal suspension, to the extent to which those properties are determined by the diffuse layer potential.

References

Grahame, D. C. (1947). *Chemical Reviews*, **41**, 441.

Hunter, R. J. (1987). *Foundations of Colloid Science*, Vol. I, p. 319. Oxford University Press.

James, R. O. and Parks, G. A. (1982). In *Surface and Colloid Science*, Vol. 12, (ed. E. Matijevic). Plenum, New York.

Overbeek, J. Th. G. (1952). In *Colloid science*, Vol. I, (ed. H. R. Kruyt), p. 162. Elsevier, Amsterdam.

MEASURING SURFACE CHARGE AND POTENTIAL

8.1 Measuring surface charge

8.1.1 Surface charge determined by titration

We have shown in §7.3.2 how the surface charge and potential of the classical silver iodide sol can be measured by titrating the colloid with a solution of one of the potential-determining ions. When applied to the many important oxide systems (for example, alumina, silica, ferric oxide, zirconia, and manganese dioxide), the technique involves titrating a known mass of material (of known surface area) with either acid or base, in the presence of a known concentration of a background (indifferent) electrolyte like KCl.

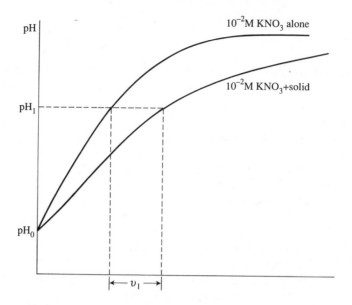

Fig. 8.1. Illustration of a potentiometric titration to establish surface charge (see text).

After each addition of base, say (of known concentration), we can calculate how many moles of OH^- ion have been added to the solution. If the pH change is then measured we know how many moles of OH^- remain in the solution. The difference is assumed to be equal to the net increase in negative charge on the surface of the colloidal oxide. There is an inherent assumption here that the only mechanism for removing OH^- from the solution is the adsorption of a hydroxyl ion or the removal of a proton from the surface:

$$-M\text{-}OH + OH^- \rightarrow -M\text{-}O^- + H_2O. \qquad (8.1)$$

If, for example, some of the oxide were to be dissolved by the addition of the base, this would give rise to a spurious result. In practice it is necessary to do a blank titration using a solution of the background electrolyte. The difference between the amounts of base required to produce a certain pH change in the oxide and that required to produce the same change in the electrolyte solution can be assumed to be equal to the number of moles of negative ion adsorbed when the pH changes from the initial to the final pH value (Fig. 8.1).

In quantitative terms, a known mass, m, of the solid oxide is added to a known volume, V, of the electrolyte of known concentration (say 10^{-3} M KCl). The initial pH, called pH_0, is noted and the sample is then titrated with, say, 10^{-2} M KOH. The volume required to achieve each solution pH might look like that shown in Fig. 8.1. (A suitable time must elapse after each addition to ensure that the surface has reached equilibrium with the solution.) If on the same diagram we now superimpose the titration curve for V ml of the 10^{-3} M electrolyte alone, then the volume v_1 corresponds to the amount of base taken up by the oxide in order to establish equilibrium with a solution of $pH = pH_1$. The net increase in (negative) surface charge, per unit area is therefore

$$-F(\Gamma_+ - \Gamma_-) = 10^{-5} v_1 F/mA \qquad (8.2)$$

for v_1 in cm^3, if A is the surface area of the solid per unit mass. Here, the Γ are the surface excesses of the H^+ and OH^- ions respectively (compare eqn (7.4)).

This calculation is repeated for all pHs greater than pH_0. The process is then repeated using an acid to titrate the oxide to pH values less than pH_0. Recall that what is measured here is the *relative* surface charge because we do not know at the beginning of the titration what the charge on the oxide is. A graph showing this relative surface charge can then be drawn up as a function of pH. Again, as in §7.3.2, the process is repeated with the same background electrolyte at other concentrations, (say 10^{-2} and 10^{-4} M). The three curves when drawn on the same scale should then look like the upper curves of Fig. 7.17. The absolute values of surface charge are obtained from these relative values by identifying the common intersection point of the three curves as the **point of zero charge** or p.z.c. of the oxide.

In practice it is often difficult to determine whether the three curves do, indeed, have a common intersection point, and without that we cannot be sure that the electrolyte is behaving as an indifferent one. Fortunately there is another way of determining the p.z.c. which will become apparent after we introduce the electrokinetic potential (see §8.6.2).

Some surfaces do not have the amphoteric nature of the oxide. They are able to be charged only in one direction, either positively or negatively. The concept of a point of zero charge is then no longer useful, although there are usually pHs at which the surface is uncharged. For example, a polymer latex made of poly-(methylmethacrylate) (PMMA) usually has a surface with some carboxyl groups on it. (These are generated in the polymerization process.) At very low pH (say below pH 4) the surface is uncharged. As the pH is increased by addition of a base, the surface becomes negatively charged by the dissociation process:

$$-COOH + OH^- \rightarrow -COO^- + H_2O. \qquad (8.3)$$

Such systems rarely produce highly charged surfaces. Even at high pH the total charge is usually less than $10\,\mu C\,cm^{-2}$.

8.1.2   *Surface charge determined by ion exchange*

The titration procedure is quite satisfactory for determining the surface charge on oxides but is not so useful for the clay minerals because much of their charge comes not from the dissociation of surface groups but from the presence of defects in the clay crystal lattice. For kaolinite, for example, the charge is negative at all pHs and it becomes more negative as the pH increases. Even at the lowest accessible pH (say pH 3) the surface charge is typically about $-5 \, \mu\text{C cm}^{-2}$ which is already comparable to that of some oxides at high pH. As the pH rises, the (negative) surface charge increases to values of around $-20 \, \mu\text{C cm}^{-2}$, which is rather higher than for most oxides. Since the clays do not have a point of zero charge it is obviously not possible to use the technique described in §8.1.1 to determine the surface charge.

The ion exchange method depends on the fact that the colloidal particles must always be neutral overall. The surface charge is therefore always balanced by an equal and opposite charge which resides in the double layer around the particle. In a naturally occurring colloid, the surface charge may be balanced by ions of many different types. The ion exchange method relies on replacing all of those counterions by ions of just one kind. The colloid is then washed with water until there are no excess ions, only the colloid and its counterions. Then the counterions are displaced by another ion and the number of displaced ions is measured (Fig. 8.2).

A common procedure involves replacing all the ions on the clay surface by ammonium ion by washing with, say, molar ammonium chloride solu-

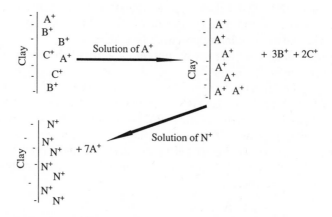

Fig. 8.2.   Principle of charge determination by ion exchange. Suspension is washed with water to remove B^+ and C^+ after step 1. Analysis of amount of A^+ in solution after step 2 gives measure of charge.

tion. The clay is then washed with water until there is no more ammonium being washed from the clay. It is then treated with NaOH which converts the NH_4^+ to ammonia and this can be distilled from the suspension and measured by titration (Exercise 8.1.1).

Exercise

8.1.1 A 5 g sample of kaolinite is saturated with ammonium ion and the excess is washed from the sample by centrifugation. The sample is then treated with an excess of sodium hydroxide. The resulting ammonia is distilled into a solution of boric acid (25 ml of 0.0107 M). The 25 ml sample of boric acid could be neutralized by 18.53 ml of a sodium hydroxide solution before the ammonia solution was distilled into it. After the distillation it required only 5.62 ml of the same sodium hydroxide for neutralization. What is the surface charge density on the clay (in millimoles of univalent charge per gram of clay)? Convert this to $\mu C\ cm^{-2}$ assuming that the clay has a surface area of $17.5\ m^2\ g^{-1}$.

8.2 The electrokinetic effects

For certain colloids, notably the silver iodide sols, and a few others mentioned in §7.3.3, it is possible to estimate the surface potential from a knowledge of the concentration of the potential-determining ion in the solution phase. That is not possible for the oxides, because they do not obey the Nernst equation, which by analogy with eqn (7.52), would be

$$\psi_0 = (RT/F)\,[pH_{pzc} - pH]$$
$$= 59.8\,[pH_{pzc} - pH]\,mV \text{ at } 25\ °C. \tag{8.4}$$

As eqn (7.59) shows, the dependence of ψ_0 on pH is a good deal more complicated for any surface which is charged as a result of the dissociation of weak acid or amphoteric groups.

But even in cases where we can estimate the surface potential reasonably accurately, it may be of little assistance in predicting the colloidal behaviour. If the solution phase contains a charged surface-active agent, which can adsorb strongly onto the particle surface, the particle's surface charge can be profoundly altered, even reversed, and that may occur with the addition of only small quantities of the surfactant. When that happens, the **effective charge** on the particle is the charge which is 'seen' by another approaching particle, because that is what will determine the interaction between the particles, and hence such things as the coagulation behaviour, or the flow of the suspension. The best estimate of that effective potential or charge is the potential or charge of the diffuse part of the double layer (Fig. 8.3).

We need, then, a measure of the charge or potential at some distance away from the surface of the particle, outside the layer of strongly adsorbed ions

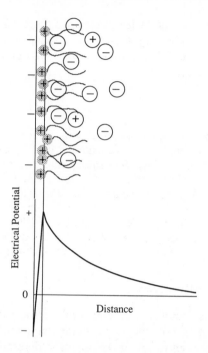

Fig. 8.3. The initial (negative) surface charge may be more than counter balanced by an oppositely charged layer of adsorbed surfactant which will make the particle appear to be positively charged.

on the surface. Ideally, we would like to be able to determine the electrostatic potential at the beginning of the diffuse part of the double layer, ψ_d. In practice we can usually only approximate that by the measurement of the **electrokinetic** or **zeta potential** (always represented by the Greek letter zeta, ζ).

There are a number of different processes by which the zeta potential can be determined. They are referred to collectively as the **electrokinetic effects**, and they occur whenever one phase is made to move with respect to the other and there is a charge at the boundary between the two phases. In the case of a colloidal suspension, that can occur when the particles settle under gravity or in a centrifuge. If the particles carry an electric charge then they will set up a measurable potential difference, called the **sedimentation potential**, when they settle (Fig. 8.4). The downward movement of the positive particle causes a slight distortion of the diffuse double layer, as the negative ions try to keep up. The result is the formation of a tiny dipole around each particle (positive end downward in the case shown in the figure). The sum of that array of dipoles is a measurable electrical potential which will make

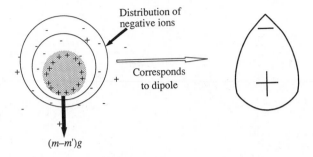

Fig. 8.4. The generation of a sedimentation potential.

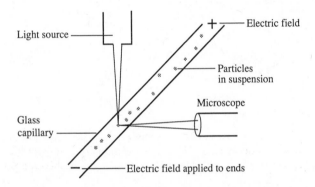

Fig. 8.5. Principle of the electrophoresis procedure. The use of a light source at right angles to the microscope enables submicroscopic particles to be seen as pinpoints of light. (This is called an ultramicroscope (Fig. 3.9).)

the bottom of the suspension positive with respect to the top. By measuring that potential difference (which is of the order of volts) we can estimate the effective charge or potential (the zeta potential) on the particles.

The same potential can be estimated by putting the suspension in an electric field and measuring the speed of the particles. That process is called **electrophoresis** and it is analogous to the transport of ions by an electric field except that, for a colloidal suspension, it is possible to follow the motion of the individual particles using a suitable microscopic technique (Fig. 8.5).

To establish the relation between the particle velocity and its electrical charge or potential it is necessary to have a clear picture of what happens at the boundary between the particle and the surrounding fluid. The layer of solvent (usually water) near the particle can always be assumed to be fixed to the particle, and normally the first few layers are so fixed. But some little distance away from the surface of the particle, the liquid begins to move with

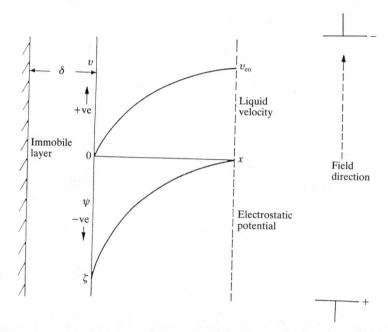

Fig. 8.6. Variation of velocity, v, and potential, ψ, in the neighbourhood of a surface in an electrokinetic process. Note that the velocity gradually increases as we move away from the surface, whilst the potential falls (that is, becomes less negative).

respect to the solid. This surface, which envelopes each particle and which marks the region where the liquid first begins to move with respect to the solid is called the **surface of shear** or, in the case of a large flat surface, the **slipping plane**. The electrostatic potential in this plane, relative to the potential in the bulk solution is called the **zeta potential** or the **electrokinetic potential** (Fig. 8.6).

Another electrokinetic process occurs when an electric field is applied to the ends of a capillary made of glass or similar material and filled with a liquid like water. The walls of the capillary will normally carry an electric charge with a size determined by the pH, and that charge will be balanced by counterions in the solution near the capillary wall. When an electric field is applied to such a system, by placing electrodes at either end of the capillary (Fig. 8.6), it will be the liquid which now moves. The counterions are moved by the electric field and, as they move, they pull the water along with them. That process is called **electro-osmosis** and the volume of liquid transported per unit time by a known electric field can be used to determine the size of the zeta potential on the capillary wall.

If the system consists of particles which are rather larger than colloidal size (say 1 μm or more) and are touching one another, with liquid in between, then applying an electric field cannot cause the particles to move. Rather the

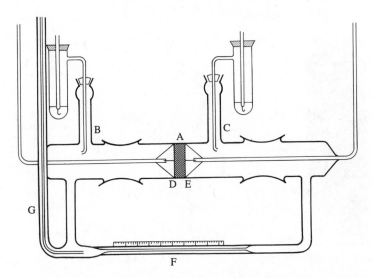

Fig. 8.7. Apparatus for measuring electro-osmosis in a porous plug, A. B and C are (reversible) electrodes through which the electrical potential is applied. The electrodes D and E can be used to measure the voltage drop across, and the conductivity of, the plug. The volume of liquid transported by electro-osmosis is measured by the scale, F, using an air bubble injected into the capillary by tube G. (After Ham and Douglas (1942) *Transactions Faraday Society*, **38**, 404.)

system can be thought of as a **porous medium** and when the electric field is applied it will be the liquid which moves in this case (Fig. 8.7). This form of electro-osmosis has been used to study the surface properties of catalysts and soil particles as well as wool and synthetic fibres like nylon and rayon.

The sedimentation potential and the electrophoresis process are closely linked. In one case the particles move and this generates an electrical potential. In the other case, a potential is applied and this causes the particles to move. Likewise if, instead of applying an electric field to a capillary, we force the fluid through it, there will be a potential generated. This is called the **streaming potential** and again it can be used to determine the zeta potential.

8.3 Calculation of the zeta potential

8.3.1 *Electro-osmosis*

The theory of the electro-osmotic effect was first given in its present form by von Smoluchowski, to whom we have already referred in Chapter 2. He

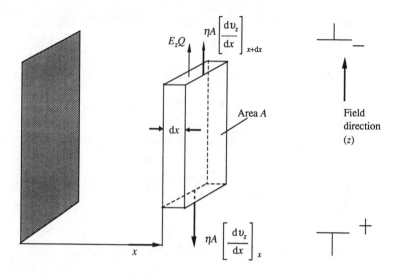

Fig. 8.8.   Forces on an element of volume of liquid of area A containing charge Q.

considered the movement of the liquid adjacent to a large, flat, charged surface, under the influence of an electric field applied parallel to the surface.
If the surface is negatively charged, there will be a net excess of positive ions
in the adjacent liquid and, as they move under the influence of the applied
field, they will draw the liquid along with them. The surface of shear in this
case may be assumed to be a plane, parallel to the solid surface and a short
distance, say δ, from it. The velocity of the liquid in the direction parallel
to the wall, v_z, rises from a value of zero in the plane of shear to a maximum value, v_{eo}, at some distance from the wall (Fig. 8.6), after which it
remains constant; v_{eo} is called the electro-osmotic velocity of the liquid.

The forces on an element of volume of the liquid are shown in Fig. 8.8.
The electrical force is equal to the product of the electric field strength and
the number of charges, Q, in the volume. This is opposed by the net frictional
force on the two sides of the slab of liquid. Therefore:

$$E_z Q = E_z \rho A\, dx = \eta A \left(\frac{dv_z}{dx}\right)_x - \eta A \left(\frac{dv_z}{dx}\right)_{x+dx}$$

$$E_z \rho\, dx = -\eta \frac{d^2 v_z}{dx^2}\, dx. \qquad (8.5)$$

Substituting for ρ from Poisson's equation (7.13) gives:

$$E_z \varepsilon_0 \varepsilon_r \left(\frac{d^2 \psi}{dx^2}\right) dx = \eta \left(\frac{d^2 v_z}{dx^2}\right) dx. \qquad (8.6)$$

This equation can be integrated from a point far from the surface up to a point in the double layer:

$$E_z \varepsilon_w \left(\frac{d\psi}{dx} \right) = \eta \left(\frac{dv_z}{dx} \right)$$

where we can assume that both derivatives are zero far from the surface. The second integration again starts from the bulk solution, where $\psi = 0$ and $v_z = v_{eo}$, and ends at the shear plane where $v_z = 0$ and $\psi = \zeta$:

$$E_z \varepsilon_w \int\limits_0^\zeta d\psi = \eta \int\limits_{v_{eo}}^0 dv_z.$$

The result is:

$$v_{eo}/E_z = u_{eo} = -\varepsilon \zeta / \eta \tag{8.7}$$

where we have assumed that both η and ε retain their bulk values all through the double layer. The quantity u_{eo} may be called the electro-osmotic mobility. The negative sign indicates that when ζ is negative, the space charge is positive so liquid flows from bottom to top in Fig. 8.8, which is the positive direction. It is important to note that the derivation does not assume anything about the arrangement of charge between the shear plane and the surface; all that is necessary is that there be no movement of the liquid (or the charge) inside that layer.

When we discuss electrophoresis in a closed capillary we will see that it is possible to measure the electro-osmotic velocity of the liquid directly, if it contains colloidal particles to act as markers. If it does not, then the liquid velocity is best determined by measuring the total volume of liquid which is transported through the capillary in unit time. The velocity profile across the capillary is as shown in Fig. 8.9. The region of varying velocity near the wall is too small to see in a microscope since it stretches only over a few times the 'double-layer thickness', $1/\kappa$ (that is, a distance of order 0.05–0.5 μm). The liquid near the wall therefore appears to move with the same velocity as that at the centre. (This is called plug flow.) The total volume transported through this open capillary is therefore

$$V = \pi r^2 v_{eo} = \pi r^2 (\varepsilon \zeta / \eta) E_z. \tag{8.8}$$

To eliminate the value of r (which may not be accurately known) we can introduce the current, i, transported by the liquid. Since the region occupied by the double layer is only a small fraction of the capillary cross-section we may assume that the liquid velocity is constant throughout and so

$$i/E_z = \pi r^2 \lambda_0 \tag{8.9}$$

where λ_0 is the electrical conductivity. Substituting in eqn (8.8) then gives

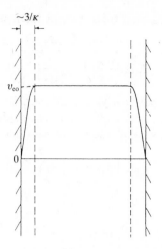

Fig. 8.9. Velocity profile of a liquid in a capillary undergoing electro-osmosis. The thickness of the region of varying velocity near the wall is greatly exaggerated.

$$V/i = \varepsilon \, \zeta/\eta \lambda_0. \tag{8.10}$$

This equation is valid only if all, or almost all, of the current is carried through the bulk liquid. If the electrolyte concentration is very low a significant part of the current may be carried through the double layers near the walls, because there is a higher charge density there. When that happens, eqn (8.9) must be replaced by

$$i/E_z = \pi r^2 \lambda_0 + 2\pi r \lambda_s$$

where λ_s is called the **specific surface conductivity**. λ_s is the conductivity of a square sheet of material of unit area and constant, though negligible, thickness measured along the length of the square; it is measured in Ω^{-1} (ohm^{-1}). Equation (8.10) then becomes

$$\frac{V}{i} = \frac{\varepsilon \, \zeta}{\eta (\lambda_0 + 2\lambda_s/r)}. \tag{8.11}$$

8.3.2 *Electrophoresis*

Smoluchowski's solution to the electrophoresis problem was simply to change the coordinate system from the solid surface to the liquid. If the liquid is regarded as fixed, the solid particle must move with a velocity equal and opposite to that of the liquid:

$$u_{\rm E} = -u_{\rm eo} = \varepsilon \zeta / \eta. \tag{8.12}$$

This relation is valid only if the double layer around the particle is very thin compared with its radius (that is, $\kappa a \gg 1$). Under those conditions, the forces imparted to the liquid by the applied electric field are transmitted to the particle as the liquid flows along its surface.

At the other extreme, when the double layer is very thick compared with the particle radius (that is, for small particles in dilute salt solutions, where $\kappa a \ll 1$) one can ignore the electrical forces acting on the double layer because they are not transmitted to the particle. The only forces the particle experiences are

(1) the direct electrical force, $f_{\rm e} = QE$, where E is the magnitude of the electric field, and Q is the particle charge; and

(2) the viscous drag of the liquid, $f_{\rm v}$, which for a spherical particle is given by the Stokes equation ((2.9) and (2.13)).

When the particle settles down to a constant velocity, these two forces are exactly balanced:

$$QE = 6\pi v a \eta \tag{8.13}$$

so that $u_{\rm E} = v/E = Q/6\pi\eta a$. Now recall that the potential, ζ, on a charged sphere is related to the charge by the expression $\zeta = Q/4\pi\varepsilon a$ and we arrive at the alternative simple relation

$$u_{\rm E} = 2\varepsilon \zeta / 3\eta. \tag{8.14}$$

This is referred to as the Hückel equation since it was proposed by Hückel as the solution of the electrophoresis problem in the early 1920s.

The question of whether to use eqn (8.12) or (8.14) was solved by Henry in the early 1930s. He showed that the two equations could be reconciled by taking account of the effect of the particle shape and size on the electric field. His equation is

$$u_{\rm E} = (2\varepsilon \zeta / 3\eta)f(\kappa a) \tag{8.15}$$

where the function f varies smoothly from 1 to 1.5 as κa varies from 0 to ∞.

The mobility is a velocity per unit field strength. It is most easily expressed in $\mu{\rm m\,s^{-1}}$ for a field strength of order $1\ {\rm V\,cm^{-1}}$ (see Exercise 8.3.3) but in SI units the mobilities of common colloidal particles are of the order of $10^{-8}\,{\rm m^2\,V^{-1}\,s^{-1}}$.

Note that for a given ζ-potential, eqns (8.12) and (8.14) show that the mobility rises by 50 per cent as the κa value increases from very small to very large values. Although Henry's equation is satisfactory if ζ is small (less than about 25 mV), as ζ rises the problem becomes a good deal more complicated. The exact relation need not concern us here. It cannot be expressed in

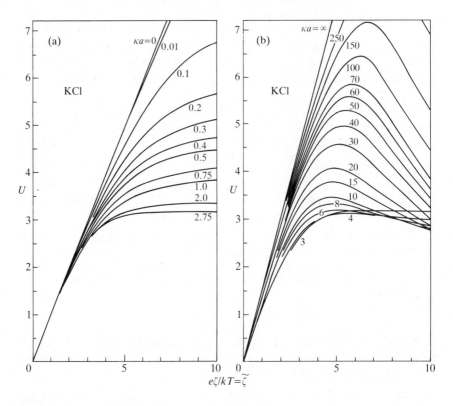

Fig. 8.10. (a) The dimensionless mobility, U, versus dimensionless ζ-potential for various (small) values of κa. (After O'Brien and White (1978), with permission.) The background electrolyte here is assumed to be KCl. The limiting value at low κa corresponds to the Hückel equation $U = \tilde{\zeta}$. (b) As for (a) but for large κa. Note that in the region $3 < \kappa a < 10$ the higher values of $\tilde{\zeta}$ have little or no effect on the mobility. The limiting value at high κa corresponds to the Smoluchowski equation $U = 3\tilde{\zeta}/2$.

analytical form and requires a sophisticated computer calculation. The result of such a calculation is shown in Figs. 8.10(a) and (b). The quantities plotted are dimensionless derivatives of the ζ-potential and the mobility. The abscissa is the reduced ζ-potential ($\tilde{\zeta} = e\zeta/kT$) which is equal to 1 when $\zeta = 25.7$ mV at 25 °C. The ordinate is the dimensionless mobility, U:

$$U = \frac{3\eta e}{2\varepsilon kT} u_e. \tag{8.16}$$

At 25 °C in water, the quantity $U = 7520 u_E$ for u_E in cm^2 V^{-1} s^{-1} (Exercise 8.3.3). In terms of these two variables, the Smoluchowski equation (8.12) becomes $U = 3\tilde{\zeta}/2$ whilst the Hückel equation is simply $U = \tilde{\zeta}$. The slope of the curves as $\tilde{\zeta}$ tends to zero is given by Henry's equation (8.15).

Exercises

8.3.1 Verify that eqn (8.10) is dimensionally correct in SI units, and show that in the commonly used c.g.s. units:

$$\frac{V(\text{cm}^3\text{s}^{-1})}{i(\text{mA})} = \frac{8.854 \times 10^{-13}\varepsilon_r \times \zeta(\text{mV})}{\eta(\text{poise}) \times \lambda_0(\Omega^{-1}\text{cm}^{-1})}.$$

8.3.2 Use the standard values for the physical constants to verify that $\tilde{\zeta} = 1$ for $\zeta = 25.7\,\text{mV}$ and that the dimensionless mobility $U = 7520\,u_E$ (for u_E in $\text{cm}^2\,\text{V}^{-1}\,\text{s}^{-1}$) at 25 °C in water.

8.3.3 Show that for particles obeying the Smoluchowski equation in water at 25 °C, $\zeta(\text{mV}) = 12.85\,u_E$ (in $\mu\text{m s}^{-1}\text{V}^{-1}$ cm.)

8.3.4 Consider the more exact expression for the potential on a sphere of radius a and charge Q surrounded by another concentric sphere of charge $-Q$ at a distance $1/\kappa$ from the first:

$$\zeta = Q/4\pi\varepsilon a - Q/\varepsilon[4\pi(a + 1/\kappa)].$$

Show that with this expression for ζ, the Hückel equation (8.14) becomes

$$u_E = \frac{2\varepsilon\zeta(1 + \kappa a)}{3\eta}.$$

(This result suggests that the Henry function for $\kappa a \ll 1$ (in eqn (8.15)) should be approximately equal to $1 + \kappa a$ but in fact that would very considerably overestimate the correction.)

8.4 Electrophoresis and electro-osmosis measurements

8.4.1 Electrophoresis

This is the most widely used of the electrokinetic procedures, especially if we include its use in the identification of proteins in complex mixtures. In that application the protein mixture is applied to one end of a column or a plate of gel (jelly) or a piece of (water saturated) paper. The electric field is then applied across the system for long enough to produce some separation of the components, on the basis of their mobilities in the medium. The rate of migration of the components can then be measured by applying a suitable stain to make the proteins visible.

As noted above, to apply the method to a colloidal suspension one must confine the suspension in a capillary (of either circular or rectangular cross-section) and apply the electric field to the ends of the capillary. That immediately introduces a complication, because the applied field causes not

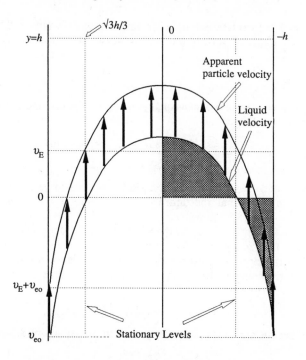

Fig. 8.11. Velocity profiles in a capillary of rectangular cross-section. The arrows represent the actual particle velocity which is constant across the cell. The lower parabola is the liquid velocity profile and the upper one is the (apparent) velocity of the particles.

only a movement of the particles (electrophoresis) but also a movement of the suspension medium (by electro-osmosis) and it is essential to take both processes into account to obtain any useful information.

The electro-osmotic flow of the fluid occurs along the capillary walls as in Fig. 8.9, but the velocity profile is very different in this case. Figure 8.9 applies to an open capillary where the fluid flows through without any impediment. In the measurement set-up shown in Fig. 8.5, the system is closed, so the fluid moving near the walls must return down the centre of the capillary. If the particles have the same sign of charge as the capillary wall, the direction of flow of the liquid at the walls will be opposite to that of the particles. (Remember that the osmotic flow occurs because of the motion of the *counterions*.) Thus, in this case, the particles near the capillary wall will be slowed down and those near the centre will be speeded up. It can be shown that the velocity profile of the liquid across the tube is parabolic (as in

Poiseuille flow, §4.2.2). Since all the particles travel at the same speed *with respect to the liquid* (assuming they have the same zeta potential) the apparent particle velocity is also parabolic (Fig. 8.11).

There is a point, at a certain distance from the capillary wall, at which the flow of liquid in each direction cancels with the opposing flow. This is the point at which one must measure the velocity of the particles, for at that level (called the **stationary level**) they are moving at their true velocity with respect to the electrodes (or the laboratory reference frame). (The shaded areas of Fig. 8.11 are equal, indicating that the osmotic flow of liquid is exactly balanced by the back flow.)

(a) Determining the position of the stationary levels in a capillary of rectangular cross-section The position of the stationary levels can be calculated from the assumption that the flow is parabolic:

$$v_p(y) = v_1(y) + v_E = ay^2 + c \tag{8.17}$$

where subscripts p and l refer to particle and liquid velocity and a and c are constants. (The linear term can be dropped by symmetry.) The liquid velocity at the wall is

$$v_1(h) = ah^2 + (c - v_E) = v_{eo} \tag{8.18}$$

where v_{eo} is the electro-osmotic velocity. Since the tube is closed:

$$2 \int_0^h v_1(y) \, dy = 0. \tag{8.19}$$

Substituting from eqn (8.17) and integrating we have

$$2 \int_0^h (ay^2 + c - v_E) \, dy = 0 \tag{8.20}$$

so that $ah^3/3 + (c - v_E)h = 0$ and hence $a = 3(v_E - c)/h^2$.

At the capillary wall, $v_1(h) = v_{eo}$ and so from eqn (8.18):

$$c = v_E - v_{eo}/2 \tag{8.21}$$

and substituting for a, c and v_E in eqn (8.17) gives

$$v_1(y) = \frac{v_{eo}}{2} \left(\frac{3y^2}{h^2} - 1 \right). \tag{8.22}$$

The stationary levels in the rectangular capillary thus occur at $y = \pm h/\sqrt{3}$. Near the walls the observed particle velocity is

$$v_p(\pm h) = v_E + v_{eo} \tag{8.23}$$

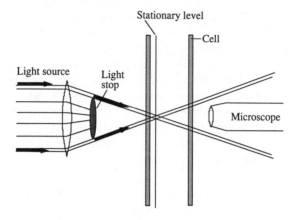

Fig. 8.12. Optical arrangement in the simple particle electrophoresis apparatus. The thickness of the cell is grossly exaggerated. It is only 1 mm across. Note that only light which is scattered by the particles can be seen in the microscope; the direct light beam does not enter the microscope objective lens.

and on the axis of the cell

$$v_p(0) = v_E - v_{eo}/2. \tag{8.24}$$

(b) Standard measurement procedure In the simple particle electro-phoresis measurement apparatus (Rank Bros, Cambridge, UK) the sample is contained in a capillary tube of total volume less than 10 ml, immersed in a thermostat bath. The electrical field is applied at the ends and the central part of the tube is a capillary of either circular or rectangular cross-section. We will confine our attention to the rectangular cell which is the easier to use. The particles are illuminated by a light source at the back of the apparatus which shines a hollow cone of light directly at the microscope objective. None of that light can pass into the objective directly. It must be scattered by a particle before it will be intercepted (Fig. 8.12). (This is called dark-field illumination and it enables one to easily distinguish particles which are less than 1 μm in size.) When viewed through the microscope, the suspended particles appear as pin-points of light (like stars) and when the field is applied they all move with essentially the same velocity.

The speed is measured by timing their progress against an eye-piece graticule (that is, a rectangular grid which sits in the eye-piece of the microscope). An improved version of the instrument incorporates a 'rotating prism' arrangement which attaches to the microscope. The prism rotates in the direction opposite to the motion of the particles and the speed of rotation can be adjusted so that the apparent particle motion is exactly counter-

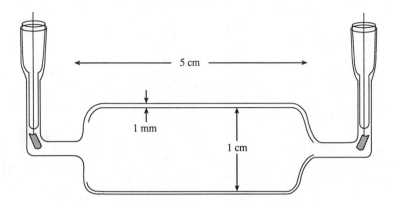

Fig. 8.13. The rectangular electrophoresis cell used by Rank Bros, Cambridge, UK.

balanced. The field of particles then appears to stop moving and the particle velocity can be read from a digital readout of the prism velocity. The device allows one to average the motion of a large number of particles at the one time.

A potential of about 60–70 V is applied to the (platinum or palladium) electrodes which are about 5–10 cm apart (Fig. 8.13). The field strength is determined by measuring the effective distance between the electrodes, by comparing the conductivity of a salt solution in the electrophoresis cell with the value obtained in a standard conductance cell. Fields of the order of $10\,V\,cm^{-1}$ are normal.

(c) More recent developments A range of more sophisticated devices is now available to measure the electrophoretic mobility of a colloidal suspension. Most are based on the use of a laser light source and some sort of data analysis by computer. The method used by Malvern Instruments is illustrated in Fig. 8.14. Two coherent beams of red light, derived by splitting the output of a low-powered helium–neon laser, are made to cross at the stationary level in the capillary cell containing the particle suspension.

At the intersection, a pattern of interference fringes† is formed. The particles are drawn across that pattern by the applied electric field and as they pass through the light and dark bands, the amount of light scattered goes through a similar fluctuation. The frequency of the fluctuations is related to the speed of the particles. The scattered light is collected by a fast photomultiplier and analysed by a digital correlator which is able to extract the frequency component due to the particle mobility. From this it constructs

† That is, alternating bands of high- and low-intensity illumination.

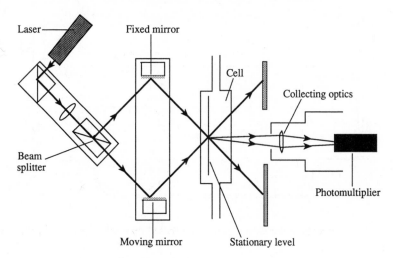

Laser

Fixed mirror

Cell

Collecting optics

Beam splitter

Photomultiplier

Moving mirror   Stationary level

Fig. 8.14. Schematic arrangement of the Malvern Zetasizer IIc system.

the mobility spectrum (that is, the frequency of particles with a given mobility). Thus one obtains a mobility *distribution* rather than just a single average value. This is sometimes of considerable importance, especially if the distribution is very broad. The sign of the mobility is determined by causing one of the mirrors to oscillate backwards and forwards; when the particles are moving in the opposite direction to the mirror they appear to move faster and so their direction of motion can be determined.

The method of **electroacoustics**, referred to in §3.7, can also be used to determine the electrophoretic mobility and, hence, the ζ-potential. The magnitude of the **colloid vibration potential**, which was described in §3.7, is proportional to the mobility of the particles, though one must recognize that it is the **dynamic** mobility rather than the usually measured **static** mobility which is important in these measurements. For particles above about 0.1 μm in size, the mobility becomes a function of frequency. The bigger particles have a larger inertia and it is this effect which can be used to measure their particle size. For smaller particles, the inertia effect is negligible (at least for frequencies below 10 MHz) but it is still possible to determine the mobility, and hence the ζ-potential. For systems in which the double layer is very thin compared to the radius, for most particles the dynamic mobility, μ_d, is given by a modified form of the Smoluchowski equation:

$$\mu_d = (\varepsilon\zeta/\eta)\,G(\omega a^2/\nu) \tag{8.25}$$

where G is the inertia correction. G depends on the frequency, ω, of the measurement, on the radius a of the particles, and on their kinematic

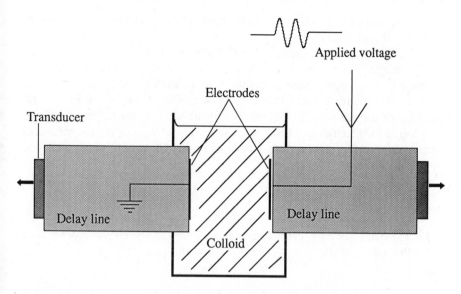

Fig. 8.15. The cell of the Matec AcoustoSizer which measures particle size and ζ-potential by measuring the magnitude and phase of the sound wave created by an applied electric field. The high-frequency (MHz) field is applied as a short pulse and the resulting sound wave moves along the delay line towards the detecting transducer. By the time it reaches there, any electronic disturbances from the applied electrical signal will have died away and the very small signal from the piezoelectric crystal (the transducer) can be measured more accurately.

viscosity $\nu(=\eta/\rho$ where ρ is the density of the solvent). In water at 25 °C, and for measuring frequencies of 1 MHz, G is near 1 for $a = 0.1$ μm and falls to about 0.5 for $a = 1$ μm. (Exercise 8.4.6).

The most recent applications of this method use the alternate effect referred to in §3.7 in which an electric field is applied to the colloid and the resulting sound wave is measured. This is called the **electrokinetic sonic amplitude** or **ESA effect**. A sketch of the measuring cell is shown in Fig. 8.15. The principal advantage of the method is that it allows measurements to be made in concentrated systems which are opaque to light and for which most other methods cannot be applied unless the sample is diluted by as much as a thousand times or more. Such dilution often changes the surface properties and makes the resulting measurements unreliable.

8.4.2 Electro-osmosis

Apart from measuring the electro-osmotic flow as an adjunct to the measurement of electrophoresis, it can also be measured independently by determining the total volume of liquid transported by a given electric current (eqns (8.10 and (8.11)).

A typical apparatus is shown schematically in Fig. 8.7. The field is applied to the plug of porous material, A, and the volume of liquid which flows through it as a consequence is measured by following the motion of an air bubble which is introduced into the lower capillary using the tube G. The packing of the material in A is rather crucial and the method has not been developed for general use, though there are some studies (like the measurement of the surface properties of wool and other fibres) for which there are few alternatives.

Exercises

8.4.1 The stationary levels in the rectangular capillary can be measured in terms of the relative depth:

$$x = (1 - y/h)/2$$

where x ranges from 0 when $y = h$ to 1 when $y = -h$. Show that they occur at

$$x_0 = 0.5 \pm (1/12)^{1/2} = 0.211 \text{ and } 0.789.$$

8.4.2 Show that the apparent particle velocity v_p can be expressed in terms of the relative depth by

$$v_p(x) = (v_E + v_{eo}) - 6v_{eo}(x - x^2).$$

If measurements are made across the entire cell, a plot of apparent velocity against the function $x - x^2$ should be linear, and from the slope and intercept the values of v_E and v_{eo} can be estimated.

8.4.3 Show that if all solid surfaces are covered with a strong adsorbate (like a protein) so that they all have the same ζ-potential, then the apparent particle velocity along the axis of the cell is $1.5 \, v_E$.

8.4.4 For a closed electrophoresis cell of circular cross-section and radius r, the liquid flow requires that

$$\int_0^r 2\pi y v_l(y) dy = 0.$$

Show that in this case $a = 2(v_E - c)/r^2$, $c = v_E - v_{eo}$, and the liquid velocity is given by

$$v_l(y) = v_{eo} \left(\frac{2y^2}{r^2} - 1 \right).$$

Thus the stationary levels in the cylindrical capillary are at $y = r/\sqrt{2} = 0.707r$.

8.4.5 Show that the velocity of the particles at the wall of the cylindrical capillary is the same as for the rectangular capillary but that on the tube axis the particle velocity is $v_E - v_{eo}$. (Compare eqns (8.23) and (8.24)). Show also that in terms of the relative depth into the capillary, $x = (1 - y/r)/2$, the particle velocity throughout the capillary is given by

$$v_p(x) = (v_E + v_{eo}) - 8v_{eo}(x - x^2).$$

8.4.6 The function G in the electroacoustic equation for the dynamic mobility is a complex number because it contains information about both the magnitude of the mobility and the phase angle between the applied field and the resulting response. The equation is

$$G(q) = \left(1 - \frac{iq[3 + 2(\Delta\rho/\rho)]}{9[1 + (1 - i)\sqrt{(q/2)}]}\right)^{-1}$$

where $q = \omega a^2/\nu$. Take $\omega = 2\pi \times 10^6$ Hz, $\Delta\rho/\rho = 2$, and $\nu = 10^{-6} \, \mathrm{m}^2 \, \mathrm{s}^{-1}$ and calculate G in the form $X + iY$ where X and Y are both real numbers, for the following values of a: 0.2 μm, 0.8 μm, 1.2 μm, 5 μm. (You will need to use the fact that $(x + iy)(x - iy) = x^2 + y^2$ to get the final form right.) What is needed for eqn (8.25) is the *magnitude* of G and that is given by $(X^2 + Y^2)^{1/2}$.

8.5   Streaming current and streaming potential

8.5.1   In single capillaries

When a liquid is forced under a hydrostatic pressure through a capillary, the charges in the mobile part of the double layer near the wall are carried towards one end. If the wall is negatively charged (as it will usually be if the capillary is made of glass or quartz) then the mobile charge is positive and so a positive current flows in the direction of the liquid flow. This is called the **streaming current**, I_s, and the accumulation of charge downstream sets up an electric field (Fig. 8.16). The field causes a current to flow back in the opposite direction and when this **conduction current**, I_c, is equal to the streaming current, a steady state is reached. The resulting electrostatic potential difference between the ends of the capillary is called the **streaming potential**. It must be measured, as a function of pressure, with a very high impedance voltmeter, so that the current flows, I_s and I_c, are not disturbed.

The relation between the streaming current and the ζ-potential is derived as follows. The flow velocity of the liquid at a distance r from the axis of the capillary (Fig. 8.17) is given by Poiseuille's equation (4.13):

$$v(r) = p(a^2 - r^2)/4\eta l \tag{8.26}$$

and the streaming current is, by definition, (compare eqn (4.14):

$$I_s = \int_0^a 2\pi r v_z(r)\rho(r)\mathrm{d}r. \tag{8.27}$$

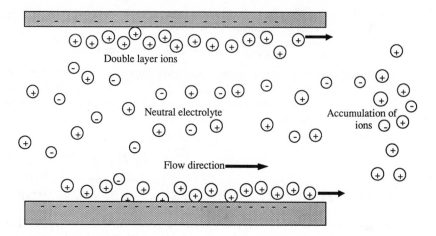

Fig. 8.16. Mechanism by which the streaming potential is generated by the accumulation of charge downstream when a fluid flows past a charged interface.

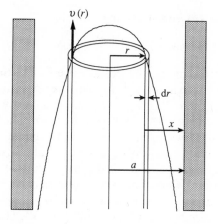

Fig. 8.17. Flow profile in a capillary in a streaming potential experiment showing a typical cylinder of fluid.

The double layer is assumed to be confined to a thin region near the wall of the capillary, so only values of r near $r = a$ are important in determining the current since the bulk of the moving liquid carries no net charge. We can substitute $r = a - x$ and, hence, when $r \approx a$:

$$v_z \approx (p/4\eta l)2a(a - r) = pax/2\eta l.$$

So

$$I_s = -\int_a^0 2\pi(a-x)\frac{pax}{2\eta l}\rho(x)\,dx \tag{8.28}$$

$$\approx -\frac{\pi a^2 p}{\eta l}\int_a^0 x\rho(x)\,dx.$$

Substituting for $\rho(x)$ from Poisson's equation (7.13) and integrating by parts gives

$$
\begin{aligned}
I_s &= \frac{\pi a^2 p}{\eta l}\int_a^0 x\varepsilon\frac{d^2\psi}{dx^2}\,dx \\
&= \frac{\pi a^2 \varepsilon p}{\eta l}\left[\left(x\frac{d\psi}{dx}\right)_{x=a}^{x=0} - \int_a^0 \frac{d\psi}{dx}\,dx\right] \\
&= -\frac{\pi a^2 \varepsilon p}{\eta l}\int_0^\zeta d\psi = -\frac{\varepsilon\zeta}{\eta}\pi a^2 p/l.
\end{aligned}
\tag{8.29}
$$

The first term in the brackets disappears because $d\psi/dx$ is zero when $x = a$.

The streaming potential, E_s, generated by this current causes a conduction current in the reverse direction given by

$$I_c = \pi a^2 E_s \lambda_0/l. \tag{8.30}$$

(Note that the field strength E_z is E_s/l in this case.) When a steady state has been established, $I_s + I_c = 0$ and so

$$E_s/p = \varepsilon\zeta/\eta\lambda_0. \tag{8.31}$$

In the usual mixed unit system:

$$\frac{E_s(mV)}{0.1 \times p(\text{dyne cm}^{-2})} = \frac{8.854 \times 10^{-12}(\text{CV}^{-1}\text{m}^{-1}) \times \varepsilon_r \times \zeta(mV)}{0.1 \times \eta(\text{poise}) \times \lambda_0(\Omega^{-1}\text{cm}^{-1}) \times 100}$$

so that, at 20 °C in water

$$\zeta = 1.055 \times 10^5 (E_s/p)\lambda_0 \tag{8.32}$$

where ζ and E_s are both measured in the same units, p is in cm Hg, and the conductivity is in $\Omega^{-1}\text{cm}^{-1}$.

Just as in the case of eqn (8.11), if we wish to take account of the **surface conduction** contribution to the back flow of charge in eqn (8.31) we must replace λ_0 by the expression $\lambda_0 + 2\lambda_s/a$. This will be important at low concentrations ($< 10^{-3}$); at higher concentrations, most of the conduction occurs through the bulk solution. The same factor then carries through to the other expressions for ζ like eqn (8.32). Usually one does not know what the surface conduction is. In that case, a simple method of correcting for it is to multiple the right-hand side of eqn (8.32) by the ratio R_{exp}/R_{calc} where

R_{exp} is the measured value of the resistance of the capillary and R_{calc} is the expected value, based on measurements made at high salt concentration, when surface conduction is negligible.

It should be noted that measurements of the streaming *current* are unaffected by the presence of the surface conduction effect. In cases where it presents a real problem, the best procedure is to measure the current instead of the potential.

8.5.2 In porous plugs

Most of the studies of streaming potential have been made on porous plugs of material, usually made up of granules with a particle size of the order of millimetres, or at least tens of micrometres. The pores in such cases are very irregular in shape but it turns out that one can apply eqn (8.31) (or the equivalent expression with surface conduction included) provided that the flow rate is not too high. It is usually assumed that if the plot of E_s as a function of applied pressure is linear† (compare eqn (8.32)), then that indicates that the pressure is not too high. Again the surface conduction can be eliminated, at least approximately, by the simple expedient of measuring the *actual conductivity* of the porous medium, rather than the value for the electrolyte solution.

8.5.3 Measurement of streaming potential and current

Measurements in single capillaries require a reversible electrode (§7.1) at either end of the capillary and a pressure head to drive the liquid through. The latter is usually provided by a low-pressure gas supply. The driving pressure can be measured with a manometer, or in more modern applications, a pressure transducer. As noted earlier, the streaming potential must be measured with a high-impedance voltmeter (more than 10^{11} Ω input impedance) like a pH meter, since the current taken from the system needs to be as near to zero as possible. Likewise, if one wishes to determine the streaming current, then that must be done with the electrodes essentially shorted out (so that there is no alternative back path for the current). In practice, the current can be estimated by connecting a small resistor (about 1000 Ω) across the electrodes and measuring the resulting voltage drop with a microvoltmeter.

For measurements on porous plugs, the plug itself is contained between two platinum gauze electrodes between which the conductivity can be measured. It is usual to determine the streaming potential using two addi-

† Sometimes the plot does not go through the origin. In such cases the slope is used. The offset is usually due to some asymmetry in the measuring electrodes.

tional electrodes just outside the platinum ones, slightly away from the plug, since the stresses in the platinum gauze interfere with the high-impedance voltage measurement.

Measurements have also been made on a parallel pair of sheets of material using an arrangement in which the electrolyte solution is forced up through a hole in the centre of the bottom sheet (Lyons *et al.* 1981). This is useful for the study of mica which has proved important in developing our understanding of the forces between colloidal particles. We will discuss that work in more detail in the next chapter.

Exercise

8.5.1 Verify the unit eqn (8.32) using the usual values for the physical properties of water.

8.6 Applications of the zeta potential

We already noted that the principal reason for determining the ζ-potential was to obtain an indication of the magnitude of the potential at the beginning of the diffuse double layer around the particle. This can then be used to estimate the effect of the particle charge on such things as aggregation behaviour, flow, sedimentation, and filtration. We will discuss some of those matters in Chapter 10. There are other, more direct, results that can be inferred directly from the ζ values themselves and these will be discussed in §8.6.2. Before doing so, however, we must examine some of the evidence that the ζ-potential does measure the electrostatic potential at the beginning of the double layer. To do that we look at the information on a model system on which a good deal of experimental work has been done.

8.6.1 Zeta potential of silver iodide surfaces

The ζ-potential of silver iodide in 0.001 M KNO$_3$ solution is shown in Fig. 8.18 along with the theoretical curve for the diffuse layer potential, ψ_d, based on a simple model of the AgI surface (Hunter 1981). The ζ values were calculated from the Henry equation to take account of the size of the particles (Osseo-Asare *et al.* 1978). The κa value is 1 in this case (Exercise 8.6.1) and Henry's correction to the Smoluchowski equation suggests that ζ is given by

$$\zeta(\text{mV}) = 20.4 \times u_E \tag{8.33}$$

where u_E is the mobility in μm s^{-1}V^{-1}cm. The ζ values for the largest mobilities were calculated using the more complete computer calculation, which gives much better agreement than the approximate (Henry) theory.

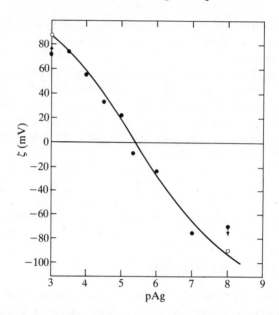

Fig. 8.18. Zeta potential of silver iodide in 0.001 M KNO_3, $\kappa a = 1$ and $T = 293\,K$. Full circles are calculated using the Henry correction. Open circles are obtained from the computer calculation. The full curve is the calculated diffuse layer potential based on the Gouy–Chapman equation for the diffuse double layer, with no specific adsorption and assuming that the inner layer capacitance, $K = 18\mu F\,cm^{-2}$.

The diffuse layer potential was estimated as follows. First it was assumed that there was no specific adsorption, so $\sigma_i = 0$ in eqn (7.42). Then from eqns (7.38) and (7.39):

$$\psi_d = \psi_0 + \sigma_d/K_i \tag{8.34}$$

and, from eqn (7.34):

$$\sigma_d = -0.1174\,c^{1/2}\sinh\left(19.46\,z\psi_d(V)\right) \tag{8.35}$$

for c in mol l^{-1}. These two equations can be solved simultaneously if we assume values for K_i and for ψ_0. The value of K_i used in the figure is $18\,\mu F\,cm^{-2}$ which is in good agreement with estimates from other adsorption measurements. ψ_0 is estimated from the Nernst equation:

$$\psi_0(V) = -0.0598\,(pAg - 5.5). \tag{7.52}$$

The agreement in this case is excellent. Unfortunately, that is rather the exception than the rule. For most systems, the ζ value tends to be rather lower in absolute magnitude than the expected ψ_d value. One should assume therefore that in most cases ζ is measured near to but slightly further from the surface than the beginning of the diffuse layer. Nevertheless, it remains our best estimate of the diffuse layer potential in most cases, and is certainly one of the most easily estimated double-layer parameters.

8.6.2 *Zeta potential of oxides and proteins*

Although they are vastly different in chemical nature, oxides and proteins are superficially similar in their ζ-potential behaviour. The surface properties of both are dominated by the effect of pH and are influenced to a lesser extent by the salt concentration. We have already noted that oxides become more negative at high pH and positive at low pH, with a cross-over point at some intermediate value. The same is true for proteins and that cross-over point can be seen in the behaviour of the ζ-potential. The mechanism is somewhat different in the two cases. For an oxide, there is usually only one ionizable surface group, but that is amphoteric so it can take up either a proton or an OH^- ion depending on the pH:

$$M\text{-}OH + H^+ \rightarrow MOH_2^+$$

or (2.25)

$$M\text{-}OH + OH^- \rightarrow MO^- + H_2O.$$

Proteins have two different surface groups, one of which can release a proton in the presence of base and the other can take up a proton if the pH is low enough:

$$R\text{-}COOH + OH^- \rightarrow R\text{-}COO^- + H_2O$$

and (8.36)

$$R'\text{-}NH_2 + H^+ \rightarrow R'\text{-}NH_3^+.$$

In each case the electrokinetic behaviour is determined by the **net charge** on the particle surface.

Figure 8.19 shows the ζ-potential of goethite ($FeO(OH)$) as a function of pH. The surface is positive at low pH where the first reaction of eqn (2.25) predominates, and is negative at high pH when the second reaction takes over. The **isoelectric point** (i.e.p.) is the point at which $\zeta = 0$. If there is no specific adsorption of the salt, then the inner layer charge density, σ_i, is zero and the surface charge is balanced only by the ions in the diffuse part of the double layer. In that case, the i.e.p. is the same as the point of zero charge (p.z.c.) of the surface (see §7.1 and Fig. 7.17). This fact is used to assist us in identifying the p.z.c. in the titration curves discussed in §8.1.1. If the titration curves at different salt concentrations appear to cross at a pH which corresponds to the i.e.p., then it can be assumed that

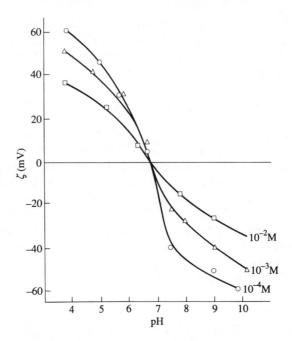

Fig. 8.19. Zeta potential as a function of pH for goethite (FeO(OH)) at different concentrations of indifferent electrolytes (10^{-2} to 10^{-4} M). (After Fuerstenau and Healy (1972), based on the work of Iwasaki *et al*.)

the i.e.p. and p.z.c. are identical, and that the salt is not being specifically adsorbed.

The main problem with this description is that when one calculates the *charge* which corresponds to the ζ-potentials shown in Fig. 8.19 it is much smaller than (often < 50 per cent of) the charge measured by direct titration (Exercise 8.6.2). For proteins it is more like 60 per cent, but there is still a marked discrepancy. There is, as yet, no completely satisfactory explanation of this phenomenon. Some of the problem may stem from the basic assumptions of the electrokinetic theory and from assumptions about the smoothness of the particle surface. Much of it can, however, be accounted for by assuming that significant amounts of *both* cations and anions are adsorbed into the inner double layer, even near the p.z.c. The reason that the p.z.c. is not much affected is that the two sorts of ions tend to cancel one another. The simple ions like K^+ and Cl^- must then be assumed to be adsorbed into the Stern layer but the phenomenon is so general and so similar for the different simple ions (on the oxide surface) that it ceases to be very specific.

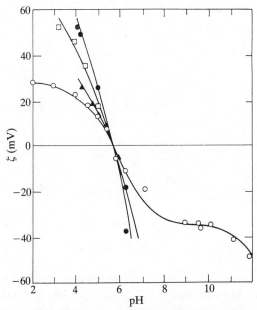

Fig. 8.20. Comparison of experimental ζ-potential data with theoretical curves calculated from a two-site model for nylon at 25 °C. Electrolyte concentration: ●, 10^{-4}; □, 10^{-3}; ▲, 5×10^{-3}; ○, 10^{-2} M. (From Rendall and Smith (1978). Hunter (1981) has more details.)

The theoretical models for some other systems have been rather more successful. Fig. 8.20 shows the excellent agreement between theory and experiment for a nylon surface using a two-site model (that is, the surface is assumed to have two sorts of groups with different acid dissociation constants). Even in this case, however, we have no independent information on the surface charge from titration measurements.

8.6.3 Specific adsorption of ions

Indifferent ions like the simple alkali metal ions and the NO_3^- ion are attracted to a charged surface by simple electrostatic forces. For an ion to be specifically adsorbed onto a surface it must get close enough to interact directly with the surface atoms. Specific adsorption of ions onto a surface is recognized by two effects: (1) it causes a shift in the p.z.c. as the ion concentration is increased, and (2) specifically adsorbed ions are able to alter the sign of the ζ-potential if they are present at high enough concentration, as we noted in §8.2.

The first effect occurs because the presence of the specifically adsorbed ions interferes with the adsorption of the potential-determining ions. The

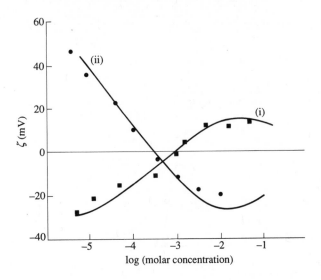

Fig. 8.21. Zeta potential as a function of electrolyte concentration for alumina at two different pH values: (i) negatively charged Al_2O_3 (at pH 10) in the presence of $BaCl_2$; (ii) positively charged Al_2O_3 (at pH 6.5) in the presence of Na_2SO_4. (See Hunter (1981).)

second effect is illustrated in Fig. 8.3. If sufficient ions are adsorbed in the inner layer they can change the sign of the diffuse double layer.

Figure 8.21 shows some data obtained for (i) a negatively charged alumina (Al_2O_3) sample at pH 10 in the presence of $BaCl_2$ and (ii) positively charged alumina (pH 6.5) in the presence of Na_2SO_4. In both cases, at concentrations around 10^{-3} M the ion of sign opposite to the initial charge is able to reverse the sign of the diffuse layer (and, hence, of the ζ-potential). The full curves in Fig. 8.21 are theoretical calculations of ζ (assumed equal to ψ_d). They are based on the idea that when $\zeta = 0$, ψ_i, the inner layer potential, is also zero. Since $\sigma_d = 0$, $\sigma_i = -\sigma_0$ from eqn (7.42). The inner layer charge can then be calculated from the surface potential using eqn (7.40) assuming some value for the inner layer capacitance (ε_i/b). The surface potential, ψ_0, was estimated from the Nernst equation which is not too far in error provided the pH is near the p.z.c. (about pH 9 for alumina). Although the ζ-potential is well described, there is again a problem with the surface charge which appears to be much smaller than that measured by titration.

8.6.4 Surfactant adsorption

As we have already noted, surfactants are used as adsorbates at the solid–liquid and the liquid–liquid interface in order to change the surface charge

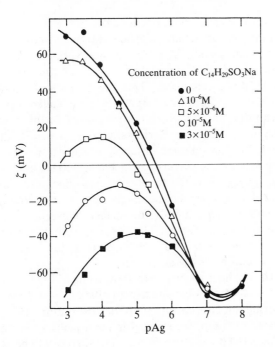

Fig. 8.22. Zeta potential as a function of pAg for various concentrations of a long-chain anionic surfactant (sodium tetradecyl sulphonate ($C_{14}H_{29}SO_3Na$). (Reproduced with permission from Osseo-Asare *et al.* (1975). Copyright American Chemical Society.)

and/or the hydrophilic/hydrophobic character of the surface. One or other of these characteristics, and often both, are important in determining such things as the floatability, rheology (flow behaviour), and filterability of a colloidal suspension or emulsion.

To understand the behaviour of different systems it is important to distinguish between solids with hydrophobic surfaces (like AgI) and those with hydrophilic surfaces, like the oxides.

On a hydrophobic surface, the hydrocarbon chain of the surfactant (Fig. 1.4) can readily displace water from the surface and so adsorption tends to begin with the hydrocarbon chain lying horizontally on the surface to make maximum contact. As the surfactant concentration increases, so too does the adsorption density, and the adsorbed hydrocarbon chains begin to interact laterally with one another and this makes adsorption more favourable. If the surface charge is not very high, and is opposite to that of the surfactant, the chains tend to remain horizontal to the surface until the solution concentration comes near to the **critical micelle concentration** (§1.4.2). Only then do the chains tend to stand vertically in order to accommodate a higher adsorption density. On the other hand, if the surface charge is high, there may be so many surfactant molecules attracted to the surface that the spaces between them become ideal for the adsorption of more surfactant. Sometimes this happens in isolated areas on the surface so that clusters of molecules are formed. (This can occur at concentrations well below the bulk c.m.c; resulting structures were called **hemimicelles** by Fuerstenau, who did much of the pioneering work in this area).

On a hydrophilic surface, the surfactant cannot so easily displace water from the surface. For low-charged surfaces, the surfactant adsorbs first with the head group against the surface, attracted by electrostatic forces. If the surface charge is very high, the number of adsorbed surfactant molecules may be so great that they are close enough together to encourage other surfactant molecules to adsorb into the spaces between them, interacting laterally by van der Waals forces. The apparent sign of the surface charge may then be changed from + to − or vice versa, but this is usually only possible near the c.m.c. of the surfactant.

Two examples will have to suffice to show some of the features of surfactant adsorption on hydrophobic and hydrophilic surfaces. Figure 8.22 shows the effect of surfactant concentration on the adsorption onto a silver iodide surface. Note that the i.e.p. for the AgI in the absence of surfactant occurs near to the anticipated p.z.c. (pAg = 5.5). The surface will be negative to the right of that point and positive to the left. The negatively charged surfactant is most obvious in its effect when the surface is positive. In fact when the surface is highly negative the surfactant has practically no effect at all. Using some plausible assumptions (Osseo-Asare *et al.* 1978) it is possible to estimate from this data, the energy of interaction of the head group with the surface (about $-3kT$) and the energy change which occurs when a CH_2 group moves from the solution to the surface. This value turns out to be about half as much as that which is involved in the micellization process. It is this sort of information, taken together with the wetting behaviour and the adsorption density, which provides evidence for the fact that on these hydrophobic surfaces the chains lie flat and expose only about half of their area to the solution.

The second illustration (Fig. 8.23) shows the importance of the chain length in determining the adsorption behaviour of a surfactant. On the hydrophilic quartz (SiO_2) surface, the ammonium acetate acts like a normal electrolyte. It can reduce the value of the ζ-potential if present in high enough concentration, but it cannot reverse the sign of charge. It is not specifically adsorbed. As the alkyl chain length is increased above C_{10} it becomes progressively easier for the positively charged surfactant to adsorb in sufficient quantity to more than balance the initial negative charge on the quartz. At the same time the initially hydrophilic surface becomes hydrophobic as the surfactant adsorbs with the head group in to the surface and the hydrocarbon chain facing the solution. This nonwetting surface can attach itself

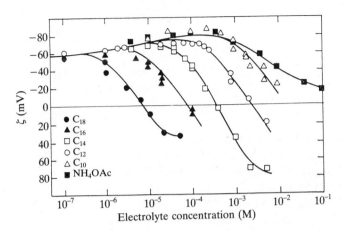

Fig. 8.23. Effect of hydrocarbon chain length on the ζ-potential of quartz in solutions of alkylammonium acetates and in solutions of ammonium acetate. (Reproduced with permission from Somasundaran *et al.* (1964). Copyright American Chemical Society.)

to an air bubble and be separated from other minerals in a flotation operation. We will discuss such possibilities in more detail in Chapter 10.

Exercises

8.6.1 Estimate the particle size of the AgI sample discussed in connection with eqn (8.33). What is the value of the Henry function f for this material?

8.6.2 Calculate the diffuse layer charge as a function of pH from the measured ζ-potentials shown in Fig. 8.19 at a concentration of 10^{-2} and 10^{-3}. Compare these with the surface charge σ_0 which for this oxide can be as high as $-10\,\mu\text{C cm}^{-2}$ at pHs well away from the p.z.c.

References

Fuerstenau, D. W. and Healy, T. W. (1972). Principles of mineral flotation. *Adsorptive bubble separation techniques*, Ch. 6. Academic, New York.

Hunter, R. J. (1981). *Zeta potential in colloid science*, Ch. 7. Academic, New York.

Lyons, J., Furlong, D. N., Homola, A., and Healy, T. W. (1981). *Australian Journal of Chemistry*, **34**, 1167–75.

O'Brien, R. W. and White, L. R. (1978). *Journal of the Chemical Society Faraday Transactions* II, **74**, 1607.

Osseo-Asare, K., Fuerstenau, D. W., and Ottewill, R. H. (1975). *Adsorption at interfaces*, American Chemical Society Symposium No. 8, p. 63. ACS, Washington, D.C. (see Hunter (1981) for more details.)

Rendall, H. M. and Smith, A. L. (1978). *Journal of the Chemical Society Faraday Transactions* I, **74**, 1179.

Somasundaram, P., Healy, T. W., and Fuerstenau, D. W. (1964). *Journal of Physical Chemistry*, **68**, 3562. (See Hunter (1981) for details.)

PARTICLE INTERACTION AND COAGULATION

9.1 Interactions between particles

In this chapter we consider some of the more important forces that come into play when two colloidal particles or two double-layer systems approach one another. Because of Brownian motion, such interactions are occurring many millions of times per second, even in quite dilute sols and, during these Brownian collisions, the double layers on the particles overlap or interpenetrate one another. The resulting interactions have a profound effect on the behaviour of the suspension, especially with respect to stability, settling, and flow. The forces can be predominantly attractive or repulsive overall, depending on the nature of the system and, in many cases, vary from attractive to repulsive and back to attractive again as the distance between the particles changes. As we noted in §2.3, the main repulsion effect often comes from electrostatic interactions. The main attraction force is due to the **London dispersion** or **van der Waals force** and will be discussed in detail in §9.2. We must now set about calculating the magnitude of these two effects.

In practice it is rather easier to deal with the **potential energy of interaction** between particles, rather than the force. The energy is a scalar quantity whereas the force is a vector, but, more importantly, we are used to the concept of an **energy barrier**, from the theory of molecular interaction and reaction kinetics, and we want to draw on that familiarity to deal with the interaction between colloidal particles and the kinetics of particle aggregation. The magnitude of the force can be estimated at any distance by looking at the slope of the potential energy curve. A positive potential energy corresponds to a repulsive force and a negative energy to an attractive force.

9.1.1 Conditions during interaction

The nature of the electrostatic interaction is influenced to some extent by how rapidly the approach occurs, since the double layers may take a significant time to adjust to the changing situation. In a highly stirred suspension the interaction will occur very rapidly and the double layers will have little chance to adjust. On the other hand, the swelling of a clayey soil as water seeps in after rain will be a much slower process where the double layer should have no difficulty in keeping pace. A collision in a quiescent suspension will be intermediate between those extremes. We should not be too surprised to find that equilibrium behaviour is often somewhat different from transient behaviour.

If the two particle surfaces approach very slowly so that equilibrium is maintained, then the surface potential should remain constant during the interaction since that is the condition of thermodynamic equilibrium. How does the potential remain constant? To answer that question fully we need to know more about what happens during the interaction. Suffice it to say at this stage that the potential-determining ions must adjust themselves to the changing circumstances as the interaction occurs.

We begin by examining the situation for small degrees of double-layer overlap because in that case the particles are not disturbed too much so the system remains close to equilibrium.

9.1.2 Overlap of two flat double layers

Even with the restriction to small degrees of overlap, the general problem is quite difficult to solve but we can get a useful insight into the solution by

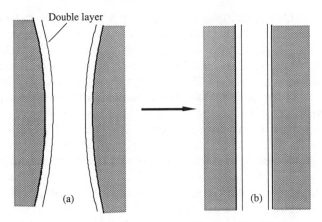

Fig. 9.1. Replacing a pair of large particles with thin double layers by a pair of flat plates.

treating the simplest possible case: that of two large particles with thin double layers approaching one another. If the double layers are sufficiently thin ($\kappa a \gg 1$) then the particles can be regarded as flat plates (Fig. 9.1). The repulsive interaction that occurs between double layers of like sign, when they begin to overlap, can be analysed by examining the **osmotic pressure** which develops due to the accumulation of ions between the plates. (If the particles approach slowly enough, then equilibrium must be maintained and so the charges on the plates must be neutralized at all stages by the ions between them; that leads to a build-up of counterions between the particles as they come near to one another.) The osmotic pressure is proportional to the local concentration, c, of ions present:

$$p = cRT = nkT \tag{9.1}$$

where c must separately sum all ion types. This pressure is very real; it is responsible for the swelling of clay minerals which can have such a profound effect on the behaviour of agricultural soils and building foundations.

The potential energy of repulsion per unit area of the plates, V_R, can be obtained by calculating the work done in forcing the plates together, from an initially infinite separation, against the opposing osmotic pressure, $\bar{p}$:

$$V_R = -\int_\infty^D \bar{p}\,\mathrm{d}D \tag{9.2}$$

where D is the final distance of separation. (The negative sign makes V_R positive since $\bar{p}$ increases with decreasing D.) Note that $\bar{p}$ is the *difference* in osmotic pressure between the solution between the plates and the solution outside.

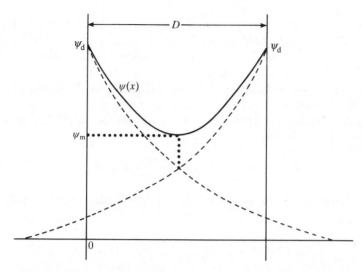

Fig. 9.2. Overlap of two diffuse double layers as two particles approach. The potential distribution in the neighbourhood of an isolated particle is shown by the broken curve. The full curve is the anticipated potential distribution for a pair of particles. ψ_m is the potential at the minimum which is in the mid-plane. The high potentials between the plates lead to high counterion concentrations and this in turn leads to a large osmotic pressure, tending to push the plates apart.

When the plates are sufficiently close so that their double layers can overlap, the electrostatic potential appears as shown in Fig. 9.2, where we assume that the 'surface' potential remains fixed at ψ_d as the particles approach. We are here assuming that only the diffuse part of the double layer needs to be considered in calculating the interaction. We will also assume that the degree of overlap is always fairly small (that is, the particles are not too close together).

Referring to Fig. 9.2 we can see that, as the particles approach, the curve describing the potential becomes shallower and so the slope of the curve at the particle surface $(d\psi/dx)$ becomes smaller (in absolute magnitude). This corresponds to a reduction in the magnitude of the surface charge (from eqn (7.32)). In other words, the plates must gradually discharge as they approach, if the potential is to remain constant. The potential-determining ions are driven from the plates and, ultimately, when the plates have come into contact, there will be no charge either on or between the plates. Only thus can the potential be kept constant. In practice, the discharge process only becomes significant when the double layers are interacting very strongly and κD is less than about 0.3, where D is the distance between the plates.

If, on the other hand, the particles move rapidly together, then there will not be time for the potential-determining ions to be desorbed from the surface and a more appropriate assumption would be that of **constant charge**. Since the difference only becomes apparent at very small separations we will not concern ourselves with it at this stage. We can learn much from the behaviour for small degrees of double-layer overlap.

The shape of the curve in Fig. 9.2 must still satisfy the Poisson–Boltzmann equation. Fortunately we can still assume that only the counterion is important in determining the behaviour and so we can use eqn (7.22), which we need to write in the form:

$$2n^0 ze \sinh \frac{ze\psi}{kT} - \varepsilon \frac{d^2\psi}{dx^2} = 0. \tag{9.3}$$

This equation can be integrated (for a fixed value of D) as before to obtain:

$$2n^0 kT \cosh \frac{ze\psi}{kT} - \frac{\varepsilon}{2}\left(\frac{d\psi}{dx}\right)^2 = p_m \text{(a constant)}. \tag{9.4}$$

Note the similarity to eqn (7.25); there is a slight difference in the boundary conditions because ψ is no longer zero far from the particle. The first term on the left is related to the osmotic pressure between the plates. That pressure varies from point to point due to the variation in ψ (and, hence, of the local ion concentration). It can be most readily evaluated at the mid-plane between the particles, because at that plane the quantity $(d\psi/dx) = 0$ and so the osmotic pressure is equal to p_m:

$$\begin{aligned} p_m &= 2n^0 kT \cosh \frac{ze\psi_m}{kT} \\ &= kT\left[n^0 \exp\left(\frac{ze\psi_m}{kT}\right) + n^0 \exp\left(-\frac{ze\psi_m}{kT}\right)\right] \\ &= kT(n_+ + n_-)_m. \end{aligned} \tag{9.5}$$

Compare this with eqn (9.1). The ion concentration at points closer to the surfaces is higher and one might think that the pressure would be higher too. But at the other points, the higher osmotic pressure is partly offset by the other term ($\frac{1}{2}\varepsilon[d\psi/dx]^2$ – called the **Maxwell stress**) so the net effect is the same throughout the whole region between the plates.

The actual force per unit area exerted on the plates is given by the difference in the osmotic pressure between the solution in the mid-plane between the plates and that outside (in the reservoir of electrolyte):

$$\begin{aligned} \bar{p} &= kT(n_+ + n_- - 2n^0) \\ &= 2n^0 kT\left(\cosh \frac{ze\psi_m}{kT} - 1\right) \\ &= 2n^0 kT(\cosh y_m - 1) \end{aligned} \tag{9.6}$$

where $y = ze\psi_m/kT$ is a dimensionless potential. If the potential in the mid-plane is small enough, the cosh term in eqn (9.6) can be expanded to give (Exercise 9.1.1)

$$\bar{p} = \tfrac{1}{2}\kappa^2\varepsilon\psi_m^2. \tag{9.7}$$

This is the expression which must be used in eqn (9.2). It is of little value yet because we need to know how ψ_m depends on the separation D. Fortunately, for the case where the overlap is small, we can assume that the potentials from the two plates simply add to one another so that:

$$\psi_m = 2\psi(x = D/2). \tag{9.8}$$

Under these conditions, eqn (7.29) is also a good approximation for ψ and so:

$$\psi_m = (8kT/ze)Z\exp(-\kappa D/2) \tag{9.9}$$

where $Z = \tanh ze\psi_d/4kT$.

The repulsion potential energy is then given by (Exercise (9.1.2))

$$V_R^\psi = -\int_\infty^D \bar{p}\,\mathrm{d}D = -\int_\infty^D 64n^0kTZ^2\exp(-\kappa D)\,\mathrm{d}D \tag{9.10}$$

$$= \frac{64n^0kTZ^2}{\kappa}\exp(-\kappa D).$$

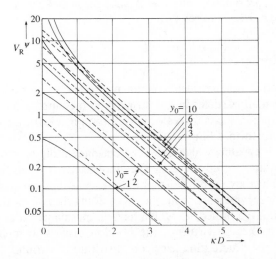

Fig. 9.3. Repulsive potential energy on a logarithmic scale against distance between the plates. Full curves: 'exact' solution; broken curves: eqn (9.10). (From Verwey and Overbeek (1948), with permission). (If κ is $10^9\,\mathrm{m}^{-1}$ then the ordinate is in units of $\mathrm{mJ\,m^{-2}}$. If κ is $10^6\,\mathrm{m}^{-1}$, the ordinate is in $\mathrm{\mu J\,m^{-2}}$.)

The superscript ψ refers to the condition of constant potential. This is one of the most widely used approximate expressions for V_R. It is also valid for conditions of constant charge since little discharge occurs if the overlap is small.

Figure 9.3 shows a comparison between the 'exact' value, calculated by complete solution of the equations, and the estimates from eqn (9.10). The relation between $\log V_R$ and D suggested by eqn (9.10) is obviously a good approximation over a wide range of D values, though the approximate equation obviously overestimates the V_R value.

Exercise

9.1.1 Use the identity $\cosh p = [\exp p + \exp(-p)]/2$ and the approximation $e^{\pm p} \approx (1 \pm p + p^2/2!)$ to derive eqn (9.7) from (9.6). What is the appropriate definition of κ for these systems?

9.1.2 Derive eqn (9.10).

9.1.3 If the electrostatic potential is small everywhere between the plates then the Debye–Hückel approximation holds:

$$\psi = \psi_d \exp(-\kappa x)]. \tag{7.28}$$

Show that under these conditions, the potential energy is given by

$$V_R = 2\varepsilon\kappa\psi_d^2 \exp(-\kappa D)$$

if the diffuse layer potential, ψ_d, is constant throughout the interaction.

9.1.4 Use the approximation $\tanh Z \approx Z$ (valid for small Z) to reconcile the result of Exercise 9.1.3 with eqn (9.10).

9.2 Attractive forces

The force of attraction between colloidal particles is a very general one which occurs between all particles in any suspension medium. It has its origins in those same van der Waals forces which are responsible for the condensation of a real gas to form a liquid.

Molecules like water, which possess a permanent dipole moment, attract one another quite strongly because the dipoles tend to align themselves, on average, so that they spend more time in orientations which are attractive rather than those which are repulsive; in so doing they lower the free energy of the system. Even molecules which have no permanent dipole are able to attract one another by virtue of the London dispersion interaction. This occurs when the fluctuating electron distribution around one atom or molecule produces a temporary dipole and that dipole induces a dipole in a neighbouring atom or molecule (Fig. 9.4). The direction of the induced dipole is always such that the two atoms attract one another.

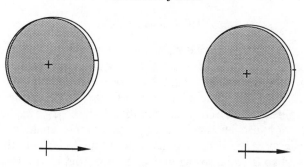

Fig. 9.4. An atom of argon has a perfectly symmetric electronic charge distribution around the central nucleus, when averaged over time. Nevertheless, on a short time-scale, the electron motions produce a temporary dipole which fluctuates in magnitude and direction at frequencies of the order of optical frequencies (10^{14} s^{-1}). The field generated by that dipole can affect the electrons in a neighbouring atom so that they tend to fluctuate to some extent in unison with one another. Since the dipoles are aligned in the same direction, they attract one another.

When this process occurs between two atoms or molecules the range of the force is of the order of a nanometre or less, but when two colloidal particles approach one another the atoms in one particle are to some extent able to interact with *all* of the atoms of the other particle and these effects are to some degree additive. The very important consequence of this partial additivity is that the force tends to be able to exert its effect over a much longer distance. It is referred to as the **'long-range van der Waals force'** or the **Hamaker force**, after the theoretician who first investigated it.

Whereas the London force between atoms falls off as the inverse sixth power of the distance of separation, we will find that this van der Waals force decreases roughly as the inverse square of the distance and so is able to exert its effects over distances of order a hundred nanometres. Thus it has a range comparable to (and in fact somewhat longer than) the electrostatic force caused by the interaction of the diffuse double layers around charged particles.

9.2.1 Attraction between molecules

A proper calculation of this interaction requires the use of quantum mechanical perturbation theory, but the general form of the result can be established by the following argument, suggested by Israelachvili (1974).

In the Bohr model of the hydrogen atom, the electron is regarded as travelling in well defined orbits about the nucleus. The orbit of smallest radius a_0 is the ground state and Bohr calculated that

$$a_0 = e^2/8\pi\varepsilon_0 h\nu \tag{9.11}$$

where e is the proton charge, ε_0 is the permittivity of free space, h is Planck's constant, and ν is a characteristic frequency (in Hz) associated with the electron's motion around the nucleus ($\nu = 3.3 \times 10^{15}\,\text{s}^{-1}$ for the Bohr hydrogen atom). (The value of a_0 given by eqn (9.11) corresponds to the maximum value in the electron density distribution $|\psi_0|^2$ in the electronic ground state of hydrogen as calculated by quantum mechanics.) The energy $-h\nu$ is the energy of the electron in its ground state (relative to the separated particles), and so is equal to the ionization potential of the H atom.

Although the H atom has no permanent dipole moment it can be regarded as having an instantaneous dipole moment, p_1, of order

$$p_1 \approx a_0 e. \tag{9.12}$$

The field of this instantaneous dipole, at a distance r from the atom, will be of the order

$$E(p_1) \approx p_1/(4\pi\varepsilon_0 r^3). \tag{9.13}$$

If a neutral atom is nearby it will therefore be polarized by this field and acquire an induced dipole moment of strength p_2:

$$p_2 = \alpha E \approx \alpha a_0 e/(4\pi\varepsilon_0 r^3) \tag{9.14}$$

where α is the atomic polarizability of the second atom. This measures the ease with which the electron distribution can be displaced (see §2.2) and is proportional to the volume of the atom (compare eqn (2.20)):

$$\alpha \approx 4\pi\varepsilon_0 a_0^3. \tag{9.15}$$

The potential energy of interaction between the dipoles p_1 and p_2 is then, using eqns (9.12) and (9.14):

$$\begin{aligned} V_A(r) &= -E(p_1)p_2 = -p_1 p_2/4\pi\varepsilon_0 r^3 \\ &\approx -\alpha a_0^2 e^2/(4\pi\varepsilon_0)^2 r^6 \end{aligned}$$

or, using eqns (9.11) and (9.15):

$$V_A(r) = -2\alpha^2 h\nu/(4\pi\varepsilon_0)^2 r^6. \tag{9.16}$$

A more exact expression can be derived from quantum mechanical considerations but the conclusion is the same: the potential energy of attraction between any two atoms, separated by a distance r, can be written

$$V_A = -B/r^6 \tag{9.17}$$

where B depends on the polarizability of the atoms. (The more polarizable the atoms are, the more easily are the electrons displaced. That produces a larger dipole and so a larger interaction force.)

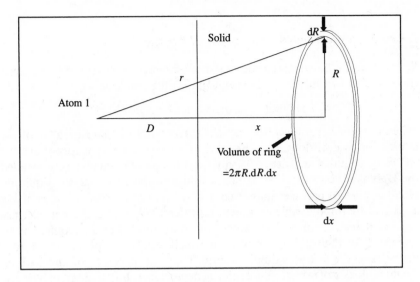

Fig. 9.5. Interaction between an atom and the surface of a solid or liquid.

9.2.2 Attraction between an atom and a wall

Now consider the interaction between an atom and a nearby surface (Fig. 9.5). To sum the interactions of all the atoms in the solid body we use the approximation of replacing the summation process by an integration:

$$V_A = - \sum_{solid} \frac{B}{r^6} dr = -B\rho \int_0^\infty dx \int_0^\infty \frac{2\pi R dR}{[(D + x)^2 + R^2]^3} \qquad (9.18)$$

where we have used Pythagoras's theorem ($r^2 = (D + x)^2 + R^2$). Equation (9.18) sums all the atoms in a ring of radius R extending out to infinity and then sums through the solid by using the variable x. The parameter ρ is the number of atoms per unit volume.

The integration process is not difficult (Exercise 9.2.1) and gives the following result:

$$V_A = -\pi B\rho/6D^3. \qquad (9.19)$$

The more gradual fall off with distance compared to the atom–atom (r^{-6}) interaction is already apparent. This is the attraction energy which is involved in the physical adsorption of a gas molecule onto a surface.

9.2.3 *Attraction between two colloidal particles*

When a similar integration process is carried out over two colloidal particles (regarded as two infinitely large flat plates) the result is

$$V_A(\text{per unit area of surface}) = -A/12\pi D^2 \qquad (9.20)$$

where D is the separation between the surfaces. Note that the range of the interaction is even longer in this case. The parameter A incorporates the values of B and ρ and depends only on the materials involved. It is called the **Hamaker constant** after the man who first did this analysis. Equation (9.20) applies to particles immersed in a vacuum but it can be shown using a general thermodynamic argument (Hunter 1987) that if the particles are immersed in a medium like water, the only modification required is to replace A for the solid by a new value which reflects the properties of both the solid and the dispersion medium. Thus the presence of the bathing medium does not affect the distance dependence but it does reduce the magnitude of the attraction. A useful approximation for the interaction between a particle of substance 1 interacting with substance 2 across a medium 3 is

$$A_{132} \approx (A_{11}^{1/2} - A_{33}^{1/2})(A_{22}^{1/2} - A_{33}^{1/2}) \qquad (9.21)$$

where, for example, the subscript 22 refers to two particles of material 2 interacting across a vacuum. It should be obvious from eqn (9.21) that the value of A for like particles interacting across a medium is always positive; that is, like particles always attract one another. For unlike particles this need not be the case.

The formulae for the van der Waals interaction for other shapes of particle can be quite complicated and we will not treat them in detail. Suffice it to quote the result for a sphere of radius a_1 interacting with one of radius a_2 with a centre-to-centre distance $H(=a_1 + a_2 + D)$. If we put $\bar{a} = (2a_1 a_2)/(a_1 + a_2)$ then in the limit when $H \ll \bar{a}$, the integration procedure yields the following result:

$$V_A(H) = -\frac{A_{12}\bar{a}}{12H}\left[1 + \frac{H}{\bar{a}}\left(1 - \frac{\bar{a}}{2(a_1 + a_2)}\right) + 2\frac{H}{\bar{a}}\ln\left(\frac{H}{\bar{a}}\right) + \ldots\right].$$
$$(9.22)$$

This equation is useful for investigating the interactions between large surfaces across a small distance, as, for instance, in the treatment of thin soap films and bubbles in a foam. For our purposes the even simpler relation for two equal spheres separated by a small distance will usually be adequate:

$$V_A = -Aa/12H \qquad (9.23)$$

where only the first term has been included.

9.2.4 Retarded van der Waals attraction

One problem with the above analysis is that it fails to take account of the fact that the interactions which are being calculated have to be propagated from one atom to another. Recall that the attraction occurs because the first dipole induces a dipole in the second atom which is oriented *in exactly the same direction as itself*. The first atom dipole generates an electric field which establishes itself through the surrounding space at the speed of light. The responding atom, under the influence of that field, becomes slightly polarized and radiates a resulting dipole field back to the first atom. It is the interaction of that field with the first atom which produces the attraction. The attraction will only be strong if the second field, when it gets back to the first atom, finds it in the same dipole orientation. If, in the meantime, the first atom has changed the orientation of its dipole significantly, the interaction will be diminished. The time taken by that whole process is determined by the finite velocity of light.

For the interaction between two atoms we do not have to concern ourselves about this effect because the force becomes insignificant before the effect comes into play. For colloid particles, however, the attraction may operate over such a large distance (of the order of a micrometre) that the finite time taken for the signal to be propagated from one atom to another is significant, even though it occurs at the speed of light. The overall result is that as the distance between the two atoms (or surfaces) increases, the force decreases more rapidly than the above equations would predict. The inverse square fall-off of eqn (9.20) gradually turns into an inverse cube law at larger

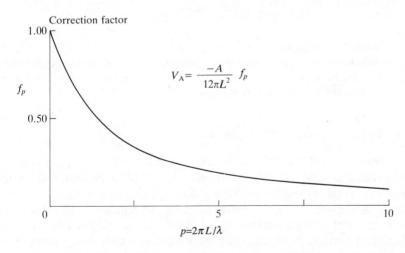

Fig. 9.6. Effect of electromagnetic retardation on the attractive interaction between two colloidal particles treated as flat plates.

distances. This is referred to as the **electromagnetic retardation effect** or sometimes as the **relativity correction**.

The theoreticians Casimir and Polder were the first to study the effect and their results have been incorporated into colloid interaction theory by Overbeek. The propagation time (r/c where c is the velocity of light) must be compared to the time taken for the electrons to undergo a significant change in arrangement ($2\pi/\omega_0 = \lambda_0/c$) where ω_0 is a characteristic electronic frequency (usually taken to be the maximum frequency of absorption of the material). Equation (9.17) must be replaced by:

$$V_A = -(B/r^6)f(p) \tag{9.24}$$

where $f(p)$ is an empirical function of the parameter $p(= 2\pi r/\lambda_0)$ (see Exercise (9.2.3)). The function f varies smoothly from about 1 to 0 as p increases from 1 to ∞. The effect of this variation in f on the calculated energy for two flat plates of infinite thickness is shown in Fig. 9.6.

Exercise

9.2.1 Establish eqn (9.19).
9.2.2 Estimate the value of A_{131} from eqn (9.21) for values of A_{11} in the range $(1\text{--}12) \times 10^{-20}$ J assuming that $A_{33} = 5.8 \times 10^{-20}$ J, the present best estimate for water. Plot the results on a log–linear scale, ignoring the values very near to $A_{33} = A_{11}$.
9.2.3 The retardation function f, in eqn (9.24) is given by Overbeek (1952) as:

$$f(p) = 1.01 - 0.14p \quad \text{for } 1 < p < 3$$

and

$$f(p) = \frac{2.45}{p} - \frac{2.04}{p^2} \quad \text{for } 3 < p < \infty.$$

Verify that these empirical equations are continuous at $p = 3$ but that the derivatives are not quite smooth at that point. Note that although $f(p)$ is equal to 0.59, for $p = 3$, the correction factor reduces the energy shown in Fig. 9.6 by a factor of about 4 at that point.

9.3 Total potential energy of interaction

The total potential energy of interaction is the sum of the repulsion and the attraction interaction. Figure 9.7 shows a plot of the dependence of the two most important influences involved in electrostatically stabilizing a colloid. The broken curves show the potential energy of repulsion (V_R) due to the electrical force whilst V_A is the van der Waals attraction. The full curves are obtained by summing each of the broken V_R curves with the V_A curve. Here we have used eqn (9.10) to estimate V_R and (9.20) for V_A.

In the expression for the attraction, it is clear that as the separation, D, between the particles decreases, the magnitude of the interaction increases without limit, so at small separations the attraction must be the dominant effect. Likewise, at very large distances, the van der Waals energy is decreasing as the inverse second (or perhaps third, if retardation is involved) power of the distance, whereas the repulsion energy is decreasing exponentially. It follows that the attraction must again dominate at large distances, because an exponential fall-off is much faster than the inverse power.

Thus, the total interaction energy is always attractive for very small and very large distances. It may be attractive for *all* distances and then the system is unstable (curve c of Fig. 9.7). Alternatively, there may be a range of distances over which the repulsion dominates. That is the situation depicted in curve a of the figure and that leads to the establishment of a barrier which can, at least to some extent, prevent the particles from coming together. The curves shown in Fig. 9.7 and the equations on which they are based lie at the heart of the theory of colloid stability which is referred to as the **DLVO theory** after the scientists, who were responsible for its development (Deryaguin, Landau, Verwey, and Overbeek).

The shape of the total potential energy curve for a sol is very important in coming to an understanding of its colloid chemical behaviour. Even the stable sol shown in curve a shows regions where the particles would attract one another (at the larger distances where the total potential energy becomes negative). The height and thickness of the barrier are also very important. It turns out that the coagulation behaviour is determined almost entirely by the height of the barrier.

To understand the importance of the barrier thickness, we need to recall how we use the potential barrier in the theory of reaction kinetics. In that case the height of the barrier is all-important and any molecule which has enough energy to surmount the barrier has a good chance of doing so. (We often have to introduce a steric factor to account for the orientation of the approaching molecules but that is usually a secondary effect.) The collision between two molecules is usually seen as a simple process in which the two approach one another on a collision path and one of two things happens. Either they collide closely enough and interact strongly enough to rearrange their bonds into a new configuration, or they make a glancing or otherwise ineffectual contact and move away without reacting.

For the colloidal particle the situation is quite different. The barrier is so thick that any particle which had enough energy to surmount it would lose that energy (in overcoming the frictional resistance of the surrounding solvent) before it could get across the barrier. Figure 9.7 shows that the barrier stretches from about $1/\kappa$ to $3/\kappa$ and at a concentration of 10^{-3} M in water, that is a distance of about 20 nm. A colloidal particle cannot simply push its way up such a hill for such a long distance. What we imagine happening is

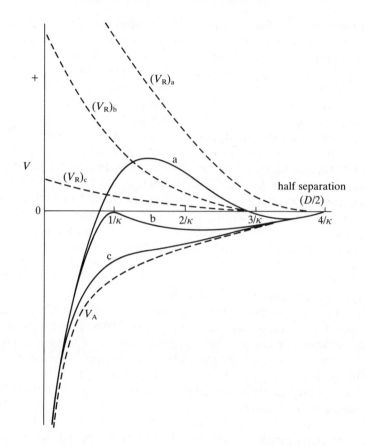

Fig. 9.7. The total potential energy of interaction, V_T, for (a) a stable, (b) a marginally stable, and (c) an unstable sol. Case (b) corresponds to the critical coagulation concentration. The curves $(V_R)_{a,b,c}$ are drawn for different values of the surface potential but approximately the same concentration of (indifferent) electrolyte.

that the particle is undergoing Brownian motion because it is being bombarded by the surrounding solvent molecules. It can either be knocked towards the second particle, or knocked in the opposite direction, and it will be more easily driven away than driven towards the second particle. Nevertheless, there is a finite, though often small, probability that it can finally fight its way over the barrier. That may only happen once in a billion collisions and that is how the energy barrier maintains the stability of the sol.

As the electrostatic potential is lowered, so that the barrier becomes lowered, a point is reached at which it just disappears; that is the critical coagulation concentration (c.c.c.) which was discussed in §2.3, Table 2.1. For still lower repulsion curves the total energy remains negative at all times, so the particles attract one another at all separations.

Notice also that there is, at long distances, a potential energy minimum (called the secondary minimum), at which the attractive forces dominate over the repulsion, even for a stable sol. Its presence is most obvious for rather large particles and it is thought to be responsible for some of the more interesting features of colloid behaviour.

Unlike the situation for molecules, it is now possible to directly measure the forces of attraction and repulsion between surfaces and to verify directly the curves shown in Fig. 9.7. We will examine those measurements shortly.

9.3.1 The Schulze–Hardy rule and the critical coagulation concentration (c.c.c.)

In discussing the data in Table 2.1 we noted the strong dependence of the critical coagulation concentration on the valency of the coagulating ion. This was referred to as the **Schulze–Hardy rule**. The approximate equations we have derived for V_R and V_A allow us to provide, for that rule, a rationalization which was one of the first and most striking results of the DLVO theory.

The c.c.c. corresponds to the point at which the maximum in the total potential energy curve just touches the horizontal axis. At this point we must have simultaneously: $V_T = 0$ and $dV_T/dD = 0$. That is:

$$V_T = -\frac{A}{12\pi d^2} + \frac{Q}{\kappa}\exp(-\kappa D) = 0$$

and (9.25)

$$\frac{dV_T}{dD} = \frac{2A}{12\pi d^3} - Q\exp(-\kappa D) = 0$$

where $Q = 64n^0kTZ^2$. Thus: $A/12\pi D^2 = (Q/\kappa)\exp(-\kappa D)$ from the first equation and $A/12\pi D^3 = (Q/2)\exp(-\kappa D)$ from the second. These are simultaneously satisfied if $\kappa D = 2$.

Substituting that value into the first equation and introducing the expression for κ^2 ($=2n^0z^2e^2/\varepsilon kT$) gives, after a little rearrangement (Exercise 9.3.2):

$$n^0 = \frac{(64 \times 24 \times \pi)^2 \times \varepsilon^3k^5T^5Z^4\exp(-4)}{2z^6e^6A^2}$$ (9.26)

where n^0 is in ions per cubic metre. Equation (9.26) shows that the c.c.c. (which is measured by n^0) is inversely proportional to the sixth power of the valency of the counterion. That is very near to the result shown in Table 2.1. The c.c.c. for valency 1, 2, and 3 for the arsenious oxide sol are: 50, 0.75, and $0.09 \, \text{mol} \, 1^{-1}$ which are in the ratio $1:0.015:0.0018$. The inverse sixth-power law would give $1:2^{-6}:3^{-6}$ which is $1:0.0156:0.0014$. This agreement is, in fact, rather too good to be true, because the assumptions on which the approximation formulae are based cannot be expected to hold for the experiments in the table. Nevertheless it is a striking illustration of the power of the theory. (We often find that scientific theories have a range of validity beyond what would be expected from the assumptions on which they are based. That is usually due to a fortuitous cancellation of errors.)

For large values of the surface potential, when $Z \approx 1$, substitution of the usual values for the variables (for water at $25 \,^\circ\text{C}$) in eqn (9.26) (Exercise 9.3.2) gives:

$$\text{c.c.c.} \, (\text{mol} \, 1^{-1}) = 88 \times 10^{-40}/z^6 A^2 \quad \text{for } A \text{ in J.} \tag{9.27}$$

A value of $50 \, \text{mmol} \, 1^{-1}$ for a $1:1$ electrolyte then corresponds to an A value of $(88 \times 10^{-40}/0.05)^{1/2} = 4.2 \times 10^{-19} \, \text{J}$. That is rather higher than the expected value (which is more like $10^{-20} \, \text{J}$) but, considering the approximations involved, the agreement is remarkable.

The most serious problem with the above analysis is the assumption that $Z \approx 1$, that is, that the surface potential is high when the system is undergoing coagulation. The general experimental observation is that coagulation normally occurs between low-potential surfaces so that assumption is invalid. We can make the more reasonable assumption of low potential and carry through the same sort of analysis (Exercises 9.3.4 and 9.3.5) and it turns out that in that case, again for water at $25 \,^\circ\text{C}$:

$$\text{c.c.c.} \, (\text{mol} \, 1^{-1}) = (3.65 \times 10^{-35}/A^2)(\psi_d^4/z^2). \tag{9.28}$$

Taking a value of $50 \, \text{mmol} \, 1^{-1}$ for the c.c.c. and the value of A suggested above $(10^{-20} \, \text{J})$ gives a value for ψ_d of $\pm 19 \, \text{mV}$ which is very reasonable. The value of ψ_d is expected to be near to the ζ-potential (§8.6) and a value of $25 \, \text{mV}$ has often been taken as the point at which a colloidal sol becomes unstable.

The earlier agreement with the Schulze–Hardy rule would then suggest that $\psi_d \propto -1/z$, so that the inverse sixth-power rule would still hold. Although this is only an empirical correlation it is certainly true that, for many systems, the magnitude of the diffuse layer potential decreases markedly as the valency of the counterion increases. This is usually assumed to occur because the higher valence counterions tend to be strongly adsorbed into the Stern layer and so drastically lower the diffuse layer potential.

Exercises

9.3.1 Redraw curve a of Fig. 9.7 onto a sheet of paper. Now draw a sketch (that is, an approximate graph) on the same diagram of the magnitude, f_R, of the force of repulsion, the force of attraction, f_A, and the net force between two particles, f_T, using the fact that $f = -dV/dD$. What is the direction of the force in each case?

9.3.2 Establish eqn (9.26) for the concentration n^0 in ions/m^3. Substitute the appropriate values into eqn (9.26) to establish the estimate of the c.c.c. in mol l^{-1} (eqn (9.27)). (Use $\varepsilon = 80\,\varepsilon_0$.)

9.3.3 Use eqn (9.27) to estimate the A values for the colloids in Table 2.1, using the data for all three valencies in each case.

9.3.4 The repulsion potential energy between two spheres of radius a for low potential and small degrees of interaction (large κH) is given by:

$$V_R = 2\pi\,\varepsilon a\psi_d{}^2\exp\left(-\kappa H\right)$$

where H is the distance of closest approach of the spheres. Use eqn (9.23) for the attraction energy and show that the c.c.c. occurs when $\kappa H = 1$. (*Hint*: use the same criterion as in the text, that is $V_T = 0$ and $dV_T/dH = 0$.) Hence show that in this case:

$$n^0 = \left(\frac{24^2\pi^2\,\varepsilon^3\psi_d{}^4}{2z^2e^2A^2}\right)kT\exp\left(-2\right)$$

where n^0 is in ions/m^3.

9.3.5 Use the result in Exercise (9.3.4) to show that, for a low-potential surface:

$$\text{c.c.c. }\left(\text{mol l}^{-1}\right) = \left(3.65\times10^{-35}/A^2\right)\psi_d{}^4/z^2.$$

9.3.6 Calculate the repulsive potential energy between two spherical particles of radius 0.5 μm with a double-layer potential of 35 mV, when the electrolyte concentration is (a) 10^{-4} M NaCl and (b) 10^{-2} M NaCl. (Use the approximation formula

$$V_R = 2\pi\,\varepsilon a\psi_d{}^2\ln\left[1 + \exp\left(-\kappa H\right)\right]$$

which applies over a wider range of κH values than the one given in Exercise 9.3.4, and take H values from 0 to 20 nm.)

9.3.7 Calculate the attraction potential energy between the particles in Exercise 9.3.6 (assuming that $A_{121} = 5\times10^{-20}$ J) as a function of H for $0 < H < 20$ nm. Combine the result with the curves for V_R found in Exercise 9.3.6 to produce curves for V_T and comment on the result.

9.3.8 The minimum which occurs at large distances in the curve for V_T calculated in Exercise 9.3.7 for a concentration of 10^{-2} M has a depth of about -1.2×10^{-19} J. Compare this to the value of kT at 25 °C and then to the gravitational potential energy of the particle if its density is 3 g cm^{-3}. How far below the reference level does the particle have to be to have the same gravitational energy as its colloidal potential energy in the well? How does this calculation relate to the development of the card-house structures shown in Fig. 2.9?

9.4 Experimental studies of the equilibrium interaction between diffuse double layers

The theory developed in §9.1–9.3 has been applied to a large number of systems over the past 50 years. More recently it has been possible to set up systems that are sufficiently well defined and controllable to provide quantitative tests of the interaction force and/or energy as a function of distance between the approaching surfaces. In this section we will examine some of the experiments conducted on double layers that are approaching sufficiently slowly to enable equilibrium to be maintained. We will also be primarily concerned with the interaction between macroscopic surfaces. Studies of the interaction between microscopic bodies (that is, colloidal particles) can be made using measurements of osmotic pressure, light scattering, or neutron scattering, among other things, and the results are in general accord with the DLVO theory. The proper discussion of those results requires the introduction of such concepts as the radial distribution function so the interpretation is not quite so straightforward as the work described here.

In the following sections we examine the behaviour of: soap films, swelling clays, and mica surfaces submerged in an electrolyte.

9.4.1 Soap films

Soap films are very easily formed by bubbling gas through a surfactant solution to produce a layer of bubbles, called a **foam**. The bubbles of gas in the foam are separated by thin, continuous films of liquid. There is a natural tendency for those films to thin down and eventually to break, under the combined influence of the gravitational force on the water in the film and the van der Waals attraction between the air bubbles on either side of the film.† Those films can only be prevented from thinning down to breaking point by the repulsive forces between the layers of surfactant molecules adsorbed at the air–solution interface.

A soap film can also be simply formed by drawing a wire frame vertically upwards through the surface of a surfactant solution. At each height, h above the solution the double-layer pressure must be balanced by the hydrostatic pressure ($h\rho g$) tending to drain the film. Hence, in a region of film 10 cm above the solution, there must be a repulsive double-layer pressure of about $0.1 \times 10^3 \times 9.8 \approx 10^3\,\mathrm{N\,m^{-2}}$. If the thickness of the

---

† It turns out that the van der Waals attraction between two gas bubbles across a film of water is the same as that between two drops of water across the same thickness of gas. We'll see how this comes about when we discuss the thinning of a soap film drawn from a solution.

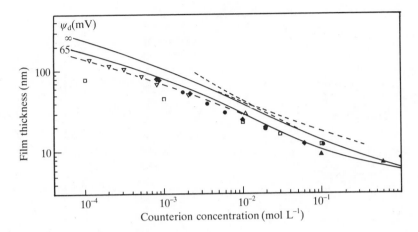

Fig. 9.8. Calculated and experimental thicknesses of soap films as a function of ionic strength. (Reproduced with permission from Lyklema and Mysels (1965). Copyright American Chemical Society.)

water layer, D, can be measured at this height we are then in a position to study the double-layer interaction.

The film thickness can be measured by the observation of reflected light from the surface of the film. Light is reflected from both faces of the film and, since there is a slight difference in the length of the path travelled by the light from the two surfaces, the two beams interfere with one another. (This is, of course, the source of the attractive colour patterns which are produced by the films, or by the films of oil on the surface of water.)

Systematic measurements on soap films were first undertaken by Derjaguin, by Scheludko, and by Lyklema and Mysels and the results of that work are shown in Fig. 9.8. The thickness at different electrolyte concentrations is shown, and compared with the expected value, after taking account of the force of gravity and the van der Waals force and adding the thickness of the adsorbed surfactant layer. As the electrolyte concentration increases, the repulsive force decreases at any particular distance and the film must thin down to a new thickness to re-establish equilibrium. The, two full lines on the figure refer to two possible surface (or double layer) potentials, either 65 mV or some large value (∞ in this context would mean any potential above about 100 mV).

Although there are still some uncertainties in this experiment there can be no doubt that the concepts of the DLVO theory give a more than adequate account of the behaviour of this sort of system. It forms the basis of our understanding of the behaviour of foams and the thinning or drainage of the liquid film from a solid when it is rapidly withdrawn from a solution. Such

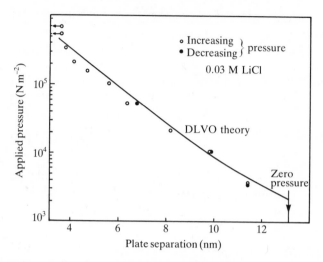

Fig. 9.9. Swelling of lithium vermiculite under pressure in 0.03 M LiCl. The theoretical curve is calculated from the total crystal charge which is almost certainly much higher than the diffuse layer charge in this system.

understanding is essential for the proper control of many industrial operations, such as the dipping of electronic components in a solution to provide them with corrosion protection.

9.4.2 Swelling of clays

The swelling of clays is important in agriculture, in civil engineering (dam, road, and building construction), and in the making of ceramics and other clay products.

In montmorillonite and vermiculite the aluminosilicate sheets are separated by water layers (§1.4.4) whose thickness varies with the concentration and type of electrolyte. In the crystalline state these extremely thin, negatively charged sheets are held together by the electrostatic forces between alternate layers of bridging cations (typically Na^+, K^+, or Ca^{2+}). The arrangement is not unlike that in an ionic solid where the cations and anions alternate in a regular way, though the pattern of charges alternates on a longer length scale in the clay system.

When the clay is placed in contact with water or a dilute electrolyte solution, some of the adsorbed counterions, being readily hydrated, tend to take up water, and may dissociate to some extent from the clay surface, to form a diffuse double layer. It is the repulsive interaction between the resulting double layers which is responsible for much of the expansion which occurs in clay mineral systems.

The earliest experiments on clay mineral swelling were done by Norrish (1954) who showed that if small crystals of montmorillonite were arranged as nearly as possible in a parallel orientation and then allowed to equilibrate with a dilute electrolyte solution $(0.01 < C$ (mol l^{-1}) < 0.25), then the separation between the sheets, d_w (in nm), was given by

$$d_w \approx 1.14(1 + C^{-1/2}).\qquad(9.29)$$

The spacing could be estimated directly from the macroscopic swelling of a known mass of clay (Exercise 9.4.1), and confirmed by low-angle X-ray measurements.

The equilibrium spacing is large compared with κ^{-1} over most of the range, so it is possible to apply the approximate equation derived for small degrees of double-layer overlap (eqn (9.10) to calculate the repulsion energy. Assuming the potential is high (so $Z^2 \approx 1$) and substituting for κ we have:

$$V_R = \frac{64}{ze}(kT)^{3/2}(\varepsilon n^0/2)^{1/2}\exp(-\kappa d_w).\qquad(9.30)$$

In the swelling region, the value of $\kappa d_w \gg 1$ and so the exponential term dominates over the square root term in this expression. We can, therefore, reduce the equation to read:

$$V_R \approx \text{const}\exp(-\kappa d_w).\qquad(9.31)$$

It seems that the facing platelets come to equilibrium at a separation distance corresponding to some fixed attractive energy $V_A = -V_R$ (a few kT per unit area of interacting plates, perhaps). Then, if V_R is constant, it follows that $d_w \propto \kappa^{-1}$ or $d_w \propto C^{-1/2}$ which was the result observed by Norrish in what he called the 'free swelling region' (eqn (9.29)).

The attraction which is involved here is thought not to be due to van der Waals forces. They are too weak to balance the repulsion at the distances involved. A more likely source is the electrostatic attraction caused by the presence of plates which are not properly aligned. We noted the need to align the plates before the experiment begins; that is rather difficult to do and some plates are probably bent or otherwise deformed in such a way that interaction can occur between the faces of some sheets and the edges of others. The positive edge charges then provide the required attraction. A few such misaligned particles would not show up in the X-ray measurements which only describe the average behaviour.

Further experiments on a related mineral called vermiculite have confirmed these early results. Vermiculite is a material similar to montmorillonite but with a better crystal structure, so that it can be better aligned. Measurements of the pressure required to force the plates into a given separation (much less than the 'free swelling' distance) allow the repulsion force to be examined alone (since it is then much larger than the van der

Waals attraction). The applied pressure decreases exponentially with plate separation with a Debye length (κ^{-1}) which agrees closely with DLVO theory (Fig. 9.9).

The magnitude of these pressures is not so easy to explain. The calculation in Fig. 9.9 assumes that the whole of the charge on the plates is dissociated into the double layer, so that it all contributes to the pressure. We would expect some of the charge to be strongly bound to the surface and this is certainly the picture suggested by the measured ζ-potential. The calculation for Fig. 9.9 also assumes that there is a layer of structured water on each crystal surface of thickness 0.5–0.6 nm. That would be consistent with the ζ-potential measurements, but only if the layer contained a significant fraction of the counterion charge. In the following section we discuss a much less ambiguous method for examining double-layer interactions.

9.4.3 The surface force apparatus

In the late 1970s a new method was developed for accurately measuring the total interaction force between molecularly smooth sheets of cleaved mica. The original apparatus was limited to operations in a vacuum but it has been developed by Israelachvili for work in aqueous electrolyte solution. The use of a crystalline layered silicate material allows the measurement of forces down to very small separations (about 0.2 nm), limited only by the smoothness of the surface, and these cleaved mica plates are the smoothest solid surfaces known. The most highly polished quartz still has hills and hollows of at least 5 nm, and often much more, over its surface.

A schematic diagram of the force-measuring apparatus is shown in Fig. 9.10. Thin, parallel-cleaved mica sheets (1–3 μm thick) are silvered on the back face and are then glued down onto curved transparent silica discs, positioned to orientate the surfaces in the geometry of crossed cylinders. White light passing through the thin sheets is multiply reflected between the silvered back surfaces so that only certain wavelengths (called FECO: fringes of equal chromatic order) are transmitted and these can be measured in a spectrometer. Analysis of the shift in wavelength from when the sheets are in contact to when they are some distance apart gives an accurate estimate of the separation (± 0.2 nm). The separation is controlled by the micrometer heads and, for the final small displacements, by the piezoelectric crystal. The force can be estimated from the deformation of the leaf spring which acts like a sensitive spring microbalance.†

Like most of the layer silicates, mica has a large negative charge (about two electronic charges per nm^2) which is balanced by potassium ions

---

† A force of 10^{-7} N is the gravitational force on a 10μg weight. The breakthrough here is not so much in the force measurement but in the accurate positioning and distance estimation.

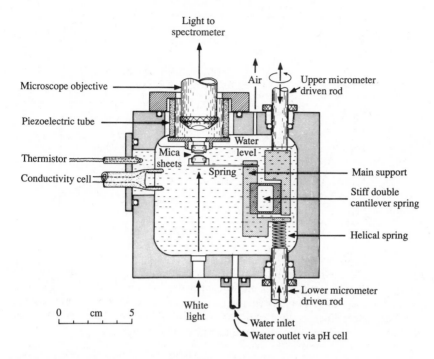

Fig. 9.10. Schematic drawing of apparatus to measure long-range forces between two crossed cylindrical sheets of mica (of thickness 1 μm and radius of curvature about 1 cm) immersed in liquid. By use of white light and multiple beam interferometry the shapes of, and separation between, the two mica surfaces may be independently measured. The separation between the two mica surfaces may be controlled by the use of two micrometer-driven rods and a piezoelectric crystal to better than 0.1 nm. (From Israelachvili and Adam (1978), with permission.)

(§1.4.4). When cleaved sheets are placed in an electrolyte solution, some of those ions dissociate to form a diffuse double layer, and it is the interaction between the double layers which can be accurately measured using the surface force apparatus. In addition, the exposed basal plane acts as a cation exchanger with the other ions in the solution. The degree of exchange will depend on the relative affinity of the different ions for the surface and the solution concentration. Thus the initial K^+ ions can be exchanged for a wide variety of other cations (like H^+, Na^+, Ca^{2+}, La^{3+}). Measurement of the interaction forces under a wide range of solution conditions can therefore give a very good test of the double-layer theory, both with respect to the repulsion and the attraction forces.

Results for the behaviour of a potassium chloride system are shown in Fig. 9.11. The agreement between theory and experiment for separations

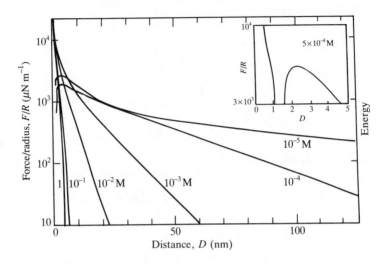

Fig. 9.11. Double-layer repulsion in the presence of potassium chloride. The Debye length, κ^{-1}, is calculated from the known concentration, but the diffuse layer potential ψ_d for each curve is adjusted for best fit. The experimental points and their associated uncertainty are covered by the thickness of the theoretical curve. The value of F/R is directly proportional to the energy of interaction.

above 5 nm is essentially exact using the theoretical estimate of the Debye length, κ^{-1}; the only adjustable parameter used is the electrostatic potential at the start of the diffuse double layer. We will discuss the behaviour at very small distances shortly.

Using the parameters established at intermediate separations, where the attraction force is unimportant, it is possible to estimate, fairly accurately, the attractive interaction at larger distances, by subtracting the calculated repulsion from the total measured force. The result is shown in Fig. 9.12, where it is clear that the force calculated using the simple Hamaker theory is adequate at spacings around 5 nm but for larger distances the theoretical estimate of the size of the attraction is too large. This is direct evidence for the importance of the retardation effect discussed in §9.2.4.

At very small separations, the simple DLVO theory would suggest that the van der Waals attraction should take over as it does in Fig. 9.7. The mica plates would then be expected to spring together. Figure 9.11 suggests that that occurs at very low salt concentrations, but above a certain value (between 10^{-3} and 10^{-4} M) the behaviour is quite different. Some additional repulsive force comes into play at distances less than about 5 nm. This region has now been studied in some detail and there appear to be two effects which are attributable to the structure of the solvent (water). The steeply rising force, evident in Fig. 9.11 for $D < 5$ nm, falls off with a characteristic

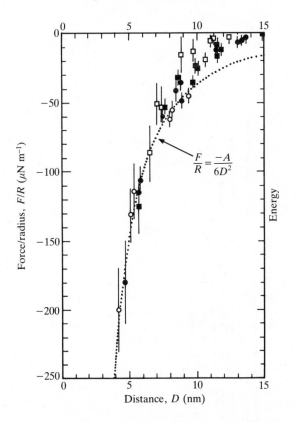

Fig. 9.12. Attractive van der Waals dispersion forces between mica surfaces measured in the region of secondary minima in various aqueous solutions. The dotted curve represents a purely nonretarded inverse square van der Waals force law. Below about 6.5 nm the forces are effectively nonretarded; above 6.5 nm they decay more rapidly, as would be expected from the theory of the retarded force. (From Israelachvili and Adams (1978), with permission.)

length of about 1 nm and is not important in more dilute solutions or at larger separations. It is assumed to be due to the adsorption, at the interface, of hydrated cations (like K^+) which cannot be squeezed out without the application of a high pressure.

The squeezing process itself can also be observed. When the behaviour at short distances is studied in detail, it reveals an oscillatory force (Fig. 9.13). The spacing of the oscillations is almost exactly equal to the diameter of a water molecule (0.25 nm), suggesting that we are here witnessing the expulsion of individual water layers from the surface. It is possible to see this effect only because of the smoothness of the surface and the fact that it is

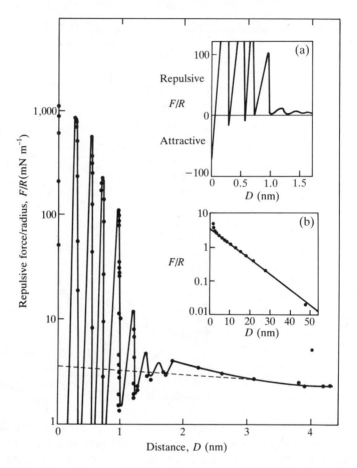

Fig. 9.13. Detailed structure of the hydration force between mica plates in 10^{-3} M KCl at very short distances. Against a solid surface the packing arrangement of the solvent molecules is revealed in an oscillating force law. (Reproduced with permission from Israelachvili and Pashley (1983). Copyright MacMillan Magazines Ltd.)

so rigid. Other comparably smooth surfaces, like the surface of an emulsion drop, or a surfactant vesicle (§1.4.2) are too easily distorted for us to see the effect. Likewise, most real surfaces are far too rough for the effect to be observed.

9.4.4 The lyotropic (or Hofmeister) series

Table 2.1 shows that there are some small differences in the coagulation behaviour of ions of the same valency. Similar differences are observed in the adsorbabilities of the monovalent anions onto the mercury surface from an electrolyte solution (Fig. 7.6). They have long been attributed to differences in the hydration energies of the different ions, because that also affects the adsorption of the ion from water. We can now observe that effect directly in the short-range forces shown in Fig. 9.13. The details of the short-range behaviour depend critically on the nature of the counterion.

There are many other situations in colloid science where the ions of a particular valency follow this same (or a very similar) order. For anions, the order of adsorbability from water is:

$$CNS^- > I^- > Br^- > Cl^- > F^- > NO_3^- > ClO_4^-$$

whilst for cations it is:

$$Cs^+ > Rb^+ > K^+ > Na^+ > Li^+.$$

These are called the **lyotropic** or **Hofmeister** series of the ions. They reflect the size of the ions and their polarizabilities. Such factors affect not only adsorption but also ion mobility. It is therefore, hardly surprising that they should have a determining effect on various aspects of both the kinetic and equilibrium behaviour of colloidal systems.

Exercise

9.4.1 In some of the experiments on clay swelling, an oriented clay in the form of a block approximately 5 mm square was used. Assume that in the 'dry' clay, each 0.97 nm thick clay plate is separated from its neighbour by a layer of water 1.12 nm thick, and that the plate has a density of 2.8 g cm^{-3}. (The density of the water is near enough to 1 g cm^{-3}.) Estimate the spacing between the plates when a block which initially weighed 52.5 mg swelled to a final mass of 402 mg. How thick was the swollen block?

9.5 Kinetics of coagulation

The rate at which a colloidal sol coagulates is one of its most important characteristics, since the properties of the sol are so profoundly affected if it is in a coagulated, rather than a dispersed, state. A large body of experimental evidence has accumulated with which to test the validity of the DLVO theory as a description of that material. Here we will only have space to introduce the central concept of the **stability ratio**, W, which measures the effectiveness of the potential barrier in preventing the particles from coagulating.

$$W = \frac{\text{Number of collisions between particles}}{\text{Number of collisions resulting in coagulation}}. \qquad (9.32)$$

We will find that values of W in excess of 10^5 are easily obtained with quite modest potential barriers $((V_T)_{max} \approx 15kT)$ and values in excess of 10^{10} are not impossible. The rate of coagulation in the absence of a potential barrier, R_f, is limited only by the rate of diffusion of the particles towards one another; that is the domain of *fast* or *rapid* coagulation. The rate of *slow* coagulation, R_s, is given by:

$$R_s = R_f / W \qquad (9.33)$$

and we will be seeking a relation between W and the height and extent of the potential barrier.

9.5.1 Rate of rapid coagulation

The rate of rapid coagulation was first calculated by von Smoluchowski and his treatment has been described by Overbeek (1952) from which this summary is taken.

We consider a central particle and calculate the number of particles that diffuse towards that particle as a result of Brownian motion. The flux, J_f, of particles whose centres pass through every sphere of radius r surrounding the central particle, and eventually come into contact with it, is given by Fick's law (§2.1.3):

$$J_f = \mathfrak{D} 4\pi r^2 \frac{dn}{dr} \qquad (9.34)$$

where n is the particle concentration and $\mathfrak{D}$ is the diffusion coefficient. This can easily be integrated (Exercise 9.5.1) to give

$$n = n_0 - J/4\pi\mathfrak{D}r \qquad (9.35)$$

where n_0 is the (bulk) particle concentration far from the central particle. The number of collisions is calculated by assuming that a particle disappears if it reaches a radius $r = 2a$ where a is the particle radius. (If the approaching particle touches the central particle it disappears because the two are then treated as a single particle). Thus $n = 0$ when $r = 2a$ and so

$$J_f = 8\pi\mathfrak{D}an_0. \qquad (9.36)$$

If the central particle is also undergoing Brownian motion the appropriate value for $\mathfrak{D}$ is $\mathfrak{D} = \mathfrak{D}_1 + \mathfrak{D}_2$ and so, if all the particles are of the same size, the initial rate of rapid coagulation is

$$-dn/dt = 1/2 \times J_f \times n_0 = 8\pi\mathfrak{D}an_0^2 = R_f. \qquad (9.37)$$

(The factor of $1/2$ must be introduced to avoid counting the collisions twice; particle X colliding with particle Y is the same as Y colliding with X.) This relation can apply only in the very early stages before there are many doublets, triplets, etc. to take into account. The rapid coagulation time is often characterized by the time, $t_{1/2}$, required for the number of particles to be reduced to half of the initial value. We can estimate that by integrating eqn (9.37), treating n_0 as a variable

$$-\int_{n_0}^{n_0/2} \frac{\mathrm{d}n}{n^2} = 8\pi\mathfrak{D}a \int_0^{t_{1/2}} \mathrm{d}t$$

$$\left(\frac{1}{n}\right)_{n_0}^{n_0/2} = 1/n_0 = 8\pi\mathfrak{D}at_{1/2}$$

so that

$$t_{1/2} = 1/8\pi\mathfrak{D}an_0. \tag{9.38}$$

Substituting for $\mathfrak{D}$ from the Einstein relation (eqn (2.12)) and using eqn (2.13) for the friction factor gives:

$$t_{1/2} = 6\pi\eta a/8\pi akTn_0 = 3\eta/4kTn_0. \tag{9.39}$$

In water at $25\,°C$ (with n_0 in particles per cm^3) this means that $t_{1/2} \approx (2 \times 10^{11}/n_0)$ seconds. Rather concentrated sols ($> 5\%$) would have n_0 values of about $10^{14}\,cm^{-3}$ and so have a rapid coagulation time in the millisecond range. The usual systems studied have values in the range of seconds to minutes.

9.5.2 Rate of slow coagulation

As we noted in §9.3 we cannot take account of the energy barrier by simply asking which particles have sufficient energy to surmount it. It is so high and so thick that a particle must negotiate the barrier by a long sequence of Brownian events, some of which will push it up the hill and some of which will push it down. The force on a particle due to the presence of the barrier is $\mathrm{d}V_T/\mathrm{d}r$ and its velocity is obtained by dividing this force by the friction factor (eqn (2.13)). The flux of particles produced by that force field is the velocity multiplied by the concentration (recall Fig. 2.3). To obtain the total flux of particles hitting the central one we must add this to the usual diffusion flux given by eqn (9.34). The total number of particles which strike the central particle at any moment is:

$$J = 4\pi r^2\left(2\mathfrak{D}\frac{\mathrm{d}n}{\mathrm{d}r} + \frac{n}{B}\frac{\mathrm{d}V_T}{\mathrm{d}r}\right). \tag{9.40}$$

Because of the mutual movement of the two particles we must replace B by $kT/2\mathfrak{D}$ (rather than $kT/\mathfrak{D}$). The solution of the resulting differential equation is (Exercise 9.5.2)

$$n = n_0 \exp(-\bar{V}) + \frac{J \exp(-\bar{V})}{8\pi\mathfrak{D}} \int\limits_{\infty}^{r} \exp \bar{V} \frac{dr}{r^2} \qquad (9.41)$$

where $\bar{V} = V_T/kT$. To satisfy the condition that $n = 0$ when $r \approx 2a$, the flux must be given by

$$J_s = \frac{8\pi\mathfrak{D}n_0}{\int_{2a}^{\infty} \exp \bar{V}(dr/r^2)} \qquad (9.42)$$

The limiting value of J when there is no potential between the particles ($V_T = 0$), except for an infinitely strong attraction when they actually make contact, is again given by

$$J_f = 16\pi\mathfrak{D}an_0. \qquad (9.43)$$

Since the rates of rapid and slow coagulation are directly proportional to the fluxes J_f and J_s, it follows from eqn (9.33) that the stability ratio is given by

$$W = R_f/R_s = J_f/J_s = 2a\int_{2a}^{\infty} \exp \bar{V} \frac{dr}{r^2}$$
$$= 2\int_{2}^{\infty} \exp\left(\frac{V}{kT}\right) \frac{ds}{s^2} \qquad (9.44)$$

where $s = r/a$. This integral must be evaluated graphically or numerically, but Verwey and Overbeek (1948) showed that W was determined almost entirely by the value of V_T at its maximum (Fig. 9.14). (This is to be expected since it enters the integrand through the exponential.) Detailed calculations suggest that for $V_{max} = 15kT$, W is of order 10^5 and for $V_{max} = 25kT$, W is about 10^9. A dilute sol, for which the rapid coagulation time is 1 s, can, therefore, be converted into a very stable sol (with a coagulation time of several months) if its double-layer potential is raised sufficiently to produce a barrier height of about $20kT$.

When W is calculated for typical values of the other parameters (ψ_0 or ψ_d, and the Hamaker constant, A) it is found that $\log_{10} W$ decreases linearly with the logarithm of the electrolyte concentration, C, over quite a wide range ($9 > \log W > 0.5$). It is also possible to provide a theoretical justification for a relationship of the form (Reerink and Overbeek 1954):

$$\log_{10} W = -k_1 \log_{10} C + k_2 \qquad (9.45)$$

where the k_i are constants. This sort of relationship has been extensively tested and confirmed experimentally (Fig. 9.15). Note that when W approaches unity ($\log W \to 0$), the system is approaching the critical coagulation concentration and any further increase in concentration can produce no further increase in the coagulation rate.

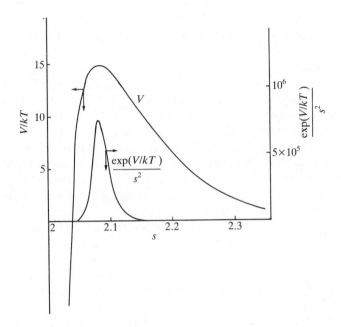

Fig. 9.14. Curves of the total potential energy of interaction, V, and of the function $s^{-2} \exp(V/kT)$ for a sol with $a = 10^{-5}$ cm, $\kappa = 10^6$ cm^{-1}, $A = 10^{-19}$ J, $\psi_d = 28.2$ mV. In this case $W \approx 5.4 \times 10^4$ (obtained by graphical integration). (From Verwey and Overbeek (1948), p. 168 with permission.)

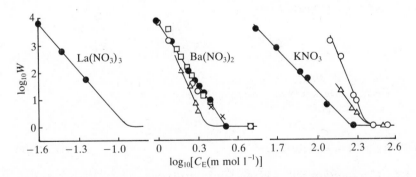

Fig. 9.15. Stability curves for various (negative) silver iodide sols at pI $= 4$. The different symbols refer to slight variations in the method of preparation; C_E is the electrolyte concentration. (From Reerink and Overbeek (1954), with permission.)

The DLVO theory of colloid stability occupies a central position in colloid science because it allows the Gouy–Chapman theory of the equilibrium double layer to be extended to situations involving double-layer overlap. This is essential, not only for the study of particle–particle interactions and coagulation, but also for such situations as sedimentation and filtration, the behaviour of electrolyte solutions in the pores of a solid (where the pore walls are normally charged), and the movement of ions through membranes.

An increasing body of evidence has developed to show that the DLVO theory is certainly valid in broad outline, and that it is able to account, almost quantitatively, for the main features of the equilibrium repulsion and attraction between smooth macroscopic surfaces and colloidal particles. Most analyses of the more complex areas of filtration, sedimentation, and electrical conduction in porous media also begin from this same fundamental standpoint. We will examine some of those situations in the next chapter.

Exercises

9.5.1 Integrate eqn (9.34) with the boundary condition $n = n_0$ when $r = \infty$ to establish eqn (9.35).

9.5.2 Under steady-state conditions, eqn (9.40) can be written:

$$\frac{J}{8\pi\mathfrak{D}} = r^2 \frac{dn}{dr} + \frac{nr^2}{kT}\frac{dV}{dr} = \text{a constant}$$

with both n and V being functions of r only. Use the differential

$$d[n\exp(V/kT)] = (n/kT)\exp(V/kT)dV + \exp(V/kT)dn$$

to show that the steady-state equation can be integrated from ∞ (where $V = 0$ and $n = n_0$) to r, to give:

$$n\exp(V/kT) - n_0 = Q\int_{\infty}^{r} \exp(V/kT)\frac{dr}{r^2}$$

where $Q(= J/8\pi\mathfrak{D})$ is constant.

Hence derive eqn (9.41) assuming that $n = 0$ at $r = 2a$.

9.5.3 Derive eqn (9.43).

9.6 Aggregate structure

9.6.1 Coagulation rate and aggregate structure

It has long been recognized that when a colloidal sol undergoes rapid coagulation, the result is a very loose aggregate in which most particles tend to be linked to only two or three other particles. The whole structure is very tenuous, and contains a great deal of entrapped solvent (usually water). The flocs settle fairly quickly, because the settling velocity is determined by the square of the radius (eqn (3.13)) and the effective aggregate radius is very large. The volume of the sediment is, however, very large and, if the system is concentrated, it may be so large that it fills the available volume so the flocs hardly settle at all. On the other hand, if the sol is undergoing slow coagulation, the flocs tend to be much more dense and though they may settle more slowly, they settle to a much smaller final volume.

This behaviour pattern is very important in many industrial situations. In the clarification of water for domestic consumption, or the cleaning of waste waters before they can be discharged into the environment, it is necessary to balance these two conflicting characteristics: to get the system to settle as quickly as possible but to produce a small final sediment volume, otherwise the cost of handling the sediment becomes very high. The problem is usually dealt with in two stages: (1) the coagulation or flocculation stage produces a clean water stream and then (2) the sediment is reduced in volume (a process called **slurry dewatering**). In large-scale operations the dewatering is done in a **thickener**.

A thickener operates by crushing the sediment under its own weight and encouraging the water to find its way up to the surface by raking the sediment with steel rods which break through the structure and form temporary channels through which the water can move upwards. In smaller-scale operations, or where the sediment is a valuable product, the suspension medium way be removed by filtration or centrifugation.

It is not difficult to see why the floc structure should depend on the rate of coagulation. When the colloid is undergoing *rapid* coagulation, there is no repulsion between the particles, and they attract one another at all separations. As soon as two particles make contact, they can stick together and they will do so. One may roll around the other as the floc moves, but if coagulation is occurring very rapidly, that will become impossible as the structure develops. The result will be a very open structure, rapid settling, and a high sedimentation volume, as is observed. In slow coagulation, the two particles approach by a diffusive process and they can only surmount the potential energy barrier if they are persistent (§9.3). The ease with which they can do so is greatly enhanced if the attractive force is increased. A particle which is moving into a position where it can interact with two particles simultaneously finds it easier to get over the barrier because the repulsive force is not much influenced by the second particle but the attractive forces are additive. Also, if the particles are nonspherical, they have to get into the optimum orientation for attractive interaction before they can get over the barrier. That will usually correspond to a well aligned, more compact arrangement. The floc therefore tends to be much more dense in its structure.

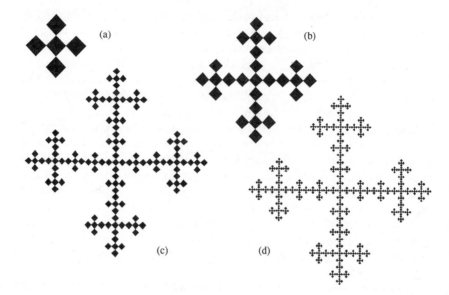

Fig. 9.16. Four stages in the construction of a simple deterministic fractal model for particle aggregates. In (a) five particles are formed in the shape of a cross. In (b) five of these crosses are joined together to form a larger cross. The process is extended in (c) and (d) in two dimensions. Each time the mass is increased by a factor of 5, the lateral extension is increased by a factor of 3. The fractal dimension of this pattern is $D = \log 5/\log 3 = 1.465$. (After Meakin (1988).)

9.6.2 *Fractal description of aggregate structure*

The loose open structure of a colloidal floc can best be described using the mathematical concept of **fractals**. This term was widely popularized by its inventor, BB Mandlebrot in the early 1980s and has since been found to be of great value in a wide variety of areas. We will do no more than touch on its significance in colloid and surface science, which has proved to be a very fertile area for its application.

Fractal geometry relates to the description of what are called **self-similar** structures, that is, structures which have the same appearance at any level of magnification. The mathematical fractals are perfectly self-similar, like the structure shown in Fig. 9.16 in which a pattern of five regular objects is replicated in two dimensions. The resulting pattern is rather like that observed in some snowflakes. Notice that if any part of the higher structure is magnified, one will see only a repetition of the same shape.

The corresponding structure in a colloidal floc is much less regular but it shares one common feature and that is that, as the structure is magnified, the appearance does not change except in respect of detail. The important parameter for describing a floc is the **fractal dimension** which is a measure of the density of the floc or the interconnectedness of the particles which make it up. Figure 9.16 shows how the dimension of the fractal in that figure is estimated. If each of the 'particles' is of equal mass, then each time the mass of the object is increased by a factor of 5, the overall size is increased by a factor of 3. The fractal dimension is given by $D = \log 5/\log 3 = 1.465$. A completely dense two-dimensional pattern would have a fractal dimension of 2, because doubling its size all round would increase the mass by a factor of 4. The actual value found for a floc in *three*-dimensional space is usually about 2.

It can be shown that light scattering measurements are able to provide a direct estimate of the fractal dimension of a growing floc. The intensity of scattering is determined by the probability of finding a particle within a certain distance of another particle and, hence, depends on the density of the floc. It turns out that the intensity, I, of the scattered light is related to the scattering vector, Q by the relation:

$$\ln I = bQ^{-D} \tag{9.46}$$

where b is a constant and D is the fractal dimension of the floc. Recall that Q is related to the wavelength, λ, and the scattering angle, θ:

$$Q = \frac{4\pi n_0}{\lambda} \sin(\theta/2). \tag{2.23}$$

Equation (9.46) is valid only for values of the scattering vector which satisfy the relation

$$a \ll Q^{-1} \ll R$$

where a is the radius of a particle and R is the radius of the aggregate. The region of validity is, however, very easily established because once the aggregates have grown to sufficient size, the plot of $\ln I$ against Q becomes strictly linear and the slope is then equal to $-D$. The most striking feature of the result is the high degree of reproducibility of the value of D for a given coagulation process carried out under repeatable conditions.

The density of the floc which is formed depends, as we noted above, on the conditions which govern the interaction between the approaching particles. Different sorts of conditions can be modelled using a computer simulation process and the resulting fractal dimension calculated. Then by comparing the measured fractal dimension with the computer results we may be able to infer which of the possible mechanisms is operating in a particular case. Typical processes for study are:

(1) single particles approaching an aggregate along a straight-line path (ballistic collision);

(2) single particles approaching an aggregate along a random walk (diffusive collision);

(3) particles requiring multiple collisions with the aggregate structure before they become attached (reaction-limited coagulation);

(4) approach of one aggregate to another either along a ballistic or a diffusive path.

These different sorts of approach pattern give different characteristic fractal dimensions ranging usually from about 1.7 to 2.2. Colloidal gold, for example, coagulated by adding pyridine (Meakin 1988) to an electrostatically stabilized sol, was found to have a D value of 1.75 by direct microscopic measurement and 1.77 by light scattering, and this is consistent with a mechanism in which clusters aggregate after diffusing towards one another (though it does not prove that that is the mechanism). The subject is still growing rapidly and currently requires much more realistic computer simulation to obtain the benefit of the very precise experimental values of fractal dimensionality.

References

Hunter, R. J. (1987). *Foundations of colloid science*, Vol. I, pp. 184–6. Oxford University Press.

Israelachvili, J. N. (1974). *Contemporary Physics*, **15**, 159.

Israelachvili, J. N. and Adams, G. E. (1978). *Journal of the Chemical Society Faraday Transactions*, I, **78**, 975.

Israelachvili, J. N. and Pashley, R. M. (1983). *Nature*, **306**, 249.

Lyklema, J. and Mysels, K. J. (1965). *Journal of the American Chemical Society*, **87**, 2539.

Meakin, P. (1988). *Advances in Colloid and Interface Science*, **28**, 250–331.

Norrish, K. (1954). *Discussions of the Faraday Society*, **18**, 120.

Overbeek, J.Th.G. (1952). In *Colloid Science*, Vol. I, p. 266, (ed. H. R. Kruyt). Elsevier, Amsterdam.

Reerink, H. and Overbeek, J.Th.G. (1954). *Discussions of the Faraday Society*, **18**, 74–84.

Verwey, E. J. W. and Overbeek, J.Th.G. (1948). *Theory of stability of lyophobic colloids*, p. 85. Elsevier, Amsterdam.

10

APPLICATIONS OF COLLOID AND
SURFACE SCIENCE

Colloidal processes and surface phenomena are important in almost every aspect of modern technology, including manufacturing, mining, and agriculture. They are also essential ingredients of the description of many biological phenomena at the cellular level and they impinge on our daily experience at every turn. For example, we have mentioned already in Chapter 5 some of the consequences of capillarity: the retention of water in soils and the way that water is drawn to the top of tall trees as a consequence of the suction pressure which develops across a curved interface (§5.5.1). It is capillarity which makes a soft towel such an effective way to dry the skin and a cloth made of the hydrophilic polymer carboxymethyl cellulose so good for wiping up spilt water. Chemists use capillary effects to separate proteins and other complex natural and synthetic products using paper or column chromatography, and in a thousand and one other ways, either consciously or unconsciously. In most situations we are unaware of the importance of capillarity in lots of the things we do and the choices we make.

In a similar way we could choose some other aspect of colloid and surface science and list the situations in which it features. Alternatively we can look at some of the specific products and processes which typify our modern environment and see what contribution is made to them by colloid and surface chemistry. From the vast array of possible subjects we will examine but a few of the more common applications.

10.1 Paper

Choosing a piece of paper for a particular purpose is more often than not a matter of wetting characteristics and capillarity. Tissue paper takes up water much more easily than writing paper, and newsprint paper looks, feels, and responds in a different way to glossy magazine paper.

10.1.1 *Making paper*

Paper, in its various forms, has been made in essentially the same way since its invention some 2000 years ago, although the modern mechanical methods have enormously increased the speed of production and the range and reproducibility of the product. Paper is made by mixing a number of colloidal, polymeric, and solution components and then allowing the colloidal suspension to flow through a narrow slit onto a wire gauze. The main component is obtained by crushing and beating, and chemically treating, wood chips to produce a pulp consisting of cellulose fibres. High-quality papers also use other fibres derived from cotton or linen† (usually added as recycled rags). The raw cellulose fibres are several millimetres in length and some tens of micrometres in diameter, but the beating process (together with the action of the chemical additives) crushes and splits them into microfibrils which are tens or hundreds of micrometres long and a few micrometres in thickness. Other colloidal components such as polymer latex particles (to bind the other components together), **fillers** (clay particles, talc, gypsum, chalk, and zinc oxide or titanium dioxide, to make the paper opaque and less absorbent), rosin (a natural polymeric material to control the wetting characteristics of the paper), and casein (a protein used as a binding agent) may be added to the mixture before it is put through the paper-making process or in the subsequent coating process. Other components, including synthetic and natural colouring agents, bleaches, adhesives, and chemicals to control the colloid stability (like aluminium sulphate, ferrous sulphate, and lime), are also required.

The colloidal suspension (at a solids concentration of about 1 per cent) is stirred thoroughly and fed to a **headbox** which is kept filled to a constant height whilst the fluid flows out through a narrow slit in the base. The fluid flows onto a continuous wire gauze or screen which is moving along under the slit at a speed of about a thousand metres or more per minute.‡ The paper pulp is a pseudoplastic material (§4.3.2.(b)) with a well defined yield value (§4.3.2.(a)) and the magnitude of the yield stress and the way in which the

---

† A cloth made from the woven or spun fibres of the flax plant.
‡ Speeds vary from less than 1000 m min^{-1} for high-quality writing paper, to about 1000–1500 for newsprint to 2000 for tissue paper.

viscosity changes with shear rate are crucial in ensuring a smooth outflow of the pulp and the correct thickness on the moving wire gauze. Those flow characteristics must be very carefully controlled and that is done by controlling the surface chemical properties of the fibres and other components by control of pH and the addition of surface-active agents which can adsorb onto the particle surfaces. Here the theoretical considerations of Chapter 4 are most relevant.

The flow behaviour is usually described by an equation like (4.23), and the primary yield value, S_0, and the viscosity $(dS/d\dot\gamma)$ are determined by the colloid chemical forces between the particles in the suspension (§9.3). Making the system more unstable tends to increase both of these quantities and the trick is to produce the desired increase in each so as to produce the best result. Although the principles are open to scientific study, the details used in any paper-making plant are closely guarded parts of the art and craft of paper making.

As the rather dilute suspension hits the wire, much of the water drains through the gauze, and the cellulose fibres form a meshwork† on which the other components are trapped. This stage of the process, when the mat is being formed on the rapidly moving wire, requires very careful control of the relevant colloidal properties. The fine fibres and the colloidal material added as filler must be retained in the mat and not allowed to drain out. Loss of material from the mat is not only wasteful, it creates a huge problem in dealing with the drainage water, most of which must be recycled. It is important that the water carries a minimum of material from one cycle to the next, since otherwise its properties cannot be adequately controlled. Modern paper plants are making increasing use of the zeta potential as a means of maintaining control over the formation of the mat. Zeta can be measured by monitoring the streaming potential (§8.5) as water runs through the mat. Normally it has a low negative value (say -5 to $-10\,\text{mV}$) and departures from that range signal the likelihood of trouble, like a future tearing of the sheet as it is pulled onto the output roller. That may involve a costly shutdown of the production line, which could feasibly be avoided by the addition of a little surfactant or an adjustment to the pH.

During its short stay on the wire, the solids content increases markedly and this is followed by a series of suction boxes which increase the solids content to around 15–20 per cent. The paper mat is then automatically transferred to a sequence of rollers which soak up the remaining excess water through layers of (wool) felt, or drive the water out with heat and pressure.

For many purposes, such as wrapping paper and newsprint, the process may then be complete, but for special papers, including good-quality writing

† The process is called 'felting' and the deposit is a form of 'felt'. The more common felt, used for making hats, is produced by a similar process from wool fibres.

or printing paper, a further coating with a clay mineral like kaolin is required to fill in the depressions in the paper surface, and improve opacity. The clay must be stuck on with a glue (either a natural polysaccharide gum or a synthetic polymer adhesive) and then the paper is again run through smooth and very heavy rollers to produce the final finish. The coating process removes any fine fibres from the surface and ensures that the effective capillary radius of the surface pores is uniform and the pores are randomly oriented. Then, when the writing or printing ink is applied to the surface, it will be taken up evenly, but not too freely, to produce a high-quality image.

When a sheet of paper is held up to the light it is possible to see the effect of the flocculated particles in the paper structure. The pore size and the surface properties of the paper particles determine the ease with which the ink can penetrate into the surface. The pores of a writing or printing paper must allow some uptake or the ink will not stick at all, as you will see if you try to write on a glossy magazine. Newsprint paper has to be able to take up ink at the enormous speeds encountered in modern high-speed printing presses, and the same is true of the hard-copy output from a computer. Producing products for these needs requires close control of the extent and rate of fluid penetration which are determined by the surface properties and the effective capillary radius since they determine the driving pressure for the flow, through eqn (5.6). The pore size is controlled by the fibre size, the amount and size of added clay and polymer latex particles, and the packing density of the fibres and fillers. The surface properties are determined chiefly by the surfactant and polymeric additives.

10.1.2 Recycling paper

Paper recycling is an important aspect of conservation, and it is relatively easy to accomplish because large amounts of used paper can be recovered from large office users and from domestic newspaper recycling. The paper-making process also lends itself to addition of used paper as a component, since it can be fairly easily returned to the condition of paper pulp. One major problem is the removal of the ink which is itself a colloidal material. Since the original printing process is designed to implant the ink permanently on the paper, it should not be surprising that getting it off again is not at all easy, especially as modern printing inks are often heat-setting plastics which can be strongly bonded to the paper surface. Redispersing the ink pigment calls for a mixture of chemical and mechanical action. In principle it is not unlike other cleaning operations, except that there is less concern about damage to the underlying fabric. The most promising method currently in use is a form of flotation similar to that described below for mineral separation. One problem is that the additives which are used to promote the flotation process can remain adsorbed on the fibres. They may then interfere when those fibres are used in a subsequent paper-making operation. The tell-tale colour of most recycled paper (and its higher than normal cost) suggest that this is a very fertile ground for the application of some good colloid chemistry.

10.2 Paints and inks

10.2.1 Paint function and composition

Paints are designed to protect and decorate the surface of metal, timber, or other structures. To do so the paint must form a smooth uniform film which is sufficiently coherent to act as a barrier to moisture and air. The earliest protective coatings were used by the ancient Egyptians and were what we would call **varnishes**. They are made by dissolving a natural polymer (called a **resin**) in an oil base. When spread in a thin film the unsaturated oil reacts with the air and undergoes an oxidative polymerization which incorporates the natural resin into a hard protective coat. Much effort was put into finding the rather rare natural resins (which are exuded by very old trees or obtained from fossilized material) and the best of them were highly prized. The paints of the time were colloidal dispersions of coloured materials (**pigments**) in water or oil with natural surfactants like proteins and polysaccharide gums being used to stabilize the colloid. Varnish was applied over the top of a plain or painted surface to provide protection. So effective was the method that the varnish on some of the sarcophagi uncovered from the tombs of the Egyptian pharaohs was still perfect after several thousand years. No doubt the dry conditions and near constant temperature prolonged their effectiveness. It is the rapid changes in temperature and the combined effects of moisture, oxygen, ultraviolet radiation, and the large variety of chemical vapours present in small concentrations in the atmosphere, coupled with the wider range of substrates, which makes the job of finding a suitable protective coating so difficult today.

It was a natural step from the discovery of varnishes to the development of paints in which the pigment is incorporated into the varnish to create what we would now call an oil-based paint or **enamel**. Much of the art and craft of paint making was secret and was centred around the bewildering variety of materials used as stabilizers for the colloidal dispersion of pigment particles, and as **drying agents** which we would now recognize as catalysts for the oxidation and polymerization of the oil film.

The introduction of synthetic resins (like nitrocellulose) in the middle of the nineteenth century made it possible to produce cheaper varnishes and quick-drying lacquers (in which a polymer is merely dissolved in the vehicle and deposits as a film when the liquid evaporates). Although the principle of polymerization was not yet understood at that time, these modified natural polymers became the basis of a large industry. The modern paint industry has moved almost entirely to the use of a variety of synthetic polymers to replace the natural oils and synthetic resins.

All paints thus have three main components: a liquid phase (called the **vehicle**), a **film-forming polymer**, and a colloidal colouring material (or **pigment**). For environmental reasons the preferred liquid these days is water, usually with some water-soluble polyalcohols (glycols) to improve the solubility of the organic components and control the drying process. The common water based paints usually use a high molecular weight ($>10^6$ daltons) polymer in the form of a latex (§1.4.5) and, as the water evaporates, the spherical polymer particles coalesce to form the protective film. Some, however, have a lower molecular weight (around 10^4–10^5) polymer which must polymerize in air before the full film strength is attained. Enamels still use an organic solvent but the pressure continues to find satisfactory water-borne substitutes.

The paint pigment has two functions: first to cover (hide) the underlying surface and then to give the desired colour. These two are often quite separate operations using separate colloidal materials. The old paints used basic lead carbonate as the hiding agent but that has been essentially entirely superseded now by the use of titanium dioxide (**titania**), in the form of particles of about 200 nm radius. Although initially higher in price, the superior hiding power and lower toxicity of titania has allowed it to become the dominant paint pigment. The colour (other than white) is then supplied as an additional, usually colloidal, component. The hiding power of titania stems from its very high refractive index (about 100 compared with values of around 5–10 for most other oxides), which gives it much greater scattering power (§2.2) for visible light. One problem with titania is that it is photoactive: that is, it can absorb light and release an electron, which is able to react with the surrounding polymer. The resulting depolymerization (or other degradation) weakens the paint film. This problem is overcome by coating the titania with a layer of alumina or silica or similar material.

10.2.2 Paint rheology

This rather complicated colloidal mixture is only the beginning, for we must now consider the other requirements for a good paint. Ease of application is a very important characteristic, especially as much paint is made for use by nonprofessionals. The paint must remain stable in the can without settling too much. If it does settle, it must be easy to redisperse by stirring. It must stick to the brush without dripping off, but must flow freely from the brush onto the surface when it is sheared by the brush-stroking action. Then it must flow sufficiently to allow the brush marks to disappear, but not so much that it will sag under the influence of gravity. Thus both the time dependence and the shear rate dependence of the viscosity and the elasticity are very important characteristics. The time effects can stretch over months or even years (for storage) to fractions of a second for the high shear stresses induced in the nozzle of a spray gun. The shear rate likewise varies from around 10^{-4} s^{-1} for the pigment settling process in the can, to about $10\,s^{-1}$ for brush-loading to about 10^5–$10^6\,s^{-1}$ for spray painting. The shear rate involved in the brushing process itself can be estimated from the typical brush speed (say $0.5\,m\,s^{-1}$) and the usual film thickness of about $50\,\mu m$. The bottom of the film, against the surface, must be stationary and the top surface is moving with the speed of the brush. The shear rate is therefore given by

$$\text{shear rate} = (0.5\,m\,s^{-1})/(50 \times 10^{-6}\,m) = 10^4\,s^{-1}$$

which is surprisingly high; it is of a similar magnitude to the maximum rate encountered in an Ostwald (capillary) viscometer (§4.2.2).

The major factor determining the viscosity is the ratio of the volume of liquid phase to the polymer and to the pigment concentrations. More subtle control is achieved by the addition of other surface-active agents, polymers, and also particulate material. According to Underwood and Jones (1991) the viscosity required for easy brushing is around 0.1–0.25 Pa s. They also suggest that a higher viscosity is required at lower shear rate to obtain good brush loading and a much higher viscosity ($> 500\,Pa\,s$) is required at very low shear rate ($10^{-4}\,s^{-1}$) to reduce settling in the can. All these requirements suggest that the paint must be highly pseudoplastic or shear-thinning (§4.3.2 (b)) and possibly thixotropic (§4.4).

A solvent-borne paint with a typical ratio of liquid to polymer to pigment (5:4:1) is almost Newtonian (§4.1). A water-based paint of the same composition is highly shear thinning. It then tends to be too viscous at low shear and not viscous enough at high shear. To get the necessary range of behaviour for the water-based paint it must first be made more dilute so that it becomes more like a Newtonian fluid. In both cases then the varying viscosity must be produced by the introduction of **thickeners**. Some thickeners are specially treated clay minerals and some are polymers. Associative thickeners, for example, are polymers which have hydrophobic and hydrophilic parts. The hydrophobic parts tend to associate with one another in water whilst the hydrophilic parts expand to interact with the water. The association is, however, very weak and is readily destroyed by the shear regime so the viscosity decreases with increasing shear rate.

Figure 10.1 shows the rheology typical of a solvent-borne paint. Note that there is a slight difference between the measured values with increasing, compared with decreasing, shear rate. That is an indication of the presence of some thixotropy. Note also the slight shear thinning of the associative thickener solution. The cellulosic thickener has a much more pronounced shear thinning characteristic.

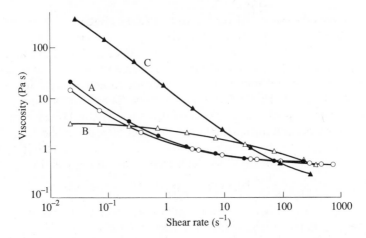

Fig. 10.1. Rheology of paints: solvent-borne paint (A) increasing (●) and decreasing (○), shear rate; latex paints formulated with associative thickeners (B), and cellulosic thickeners (C). (From Underwood and Jones (1991), with permission.)

Another important characteristic is the way in which the viscosity changes with increase in the particle concentration. This is especially true of waterborne paints which tend to dry fairly quickly. As they dry, the increase in the volume fraction of solids causes a further increase in viscosity as would be expected from eqn (4.20). At still higher volume fractions, the behaviour may change from shear thinning to shear thickening (Fig. 10.2) since shear thickening (or dilatant) behaviour is common for very concentrated systems. This can result in 'sticky', partially dried, paint which cannot be worked for long enough to produce smooth overlap into the next brush-load of paint.

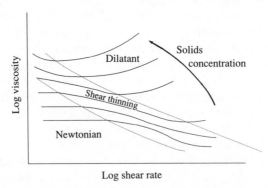

Fig. 10.2. Schematic plot showing the non-Newtonian behaviour of a paint, generalized for stable dispersions as a function of solids concentration. (From Underwood and Jones (1991).)

10.2.3 Ink

The basic characteristics of inks are very similar to those of paints. The simplest hand-writing inks, such as those used in fountain pens, consist essentially of a dye dissolved in a solvent, either a water base with some glycol or an organic solvent. In practice, however, there are a variety of other criteria which a good ink must satisfy: it must run freely without clogging, not deposit a sediment, not allow the development of fungal or bacterial growth on prolonged storage, not corrode the steel writing nib and metal parts nor swell the plastic parts of the pen. It must also produce a clear image, so the pigment must not run into the fine capillaries adjacent to the written mark. It must also dry reasonably quickly and not penetrate through to the opposite side of the paper. Again, in order to satisfy all those requirements (and more) the ink maker must exercise fine control over the colloid and surface chemical properties of the components, since all of these effects depend on the intimate interaction between the colloidal particles and dissolved substances in the ink and the surfaces with which they come in contact. Most of those problems were solved over the years by empirical (trial and error) procedures, but the demands of modern high-speed printing have made it necessary to replace such methods by a more scientifically based approach.

The widely used ball-point pen ink is a more interesting material from the colloid viewpoint. A ball-point pen has a reservoir of ink which is of similar composition to printing ink. The ball is about 1 mm in diameter and it is held in such a way that it can rotate freely when it is in contact with the writing surface without falling out (Fig. 10.3). The reservoir contains from 0.5 to 1.5 cm³ of ink in the form of a semi-solid gel. The ink has a significant yield point so that it will not run from the end of the pen until the ball is moved across the paper. The motion of the ball produces a shearing action on the ink and this is sufficient to exceed the yield point so that the ink flows over the ball and onto the paper. As soon as the shearing action ceases, the ink again becomes stationary and then fairly quickly dries, either by evaporation

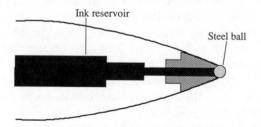

Ink reservoir

Steel ball

Fig. 10.3. Mechanism of the ball-point pen (highly schematic).

of the solvent or by the rapid infiltration of the solvent into the capillary pores of the paper, leaving the pigment behind. The ink composition and properties are similar to that used in printing inks but the yield value is very pronounced. They are nonideal Bingham plastic materials (Fig. 4.7, curve 5). The yield value is provided by including in the formulation some clay mineral particles (usually an organically modified montmorillonite) which form a loose structure which holds the ink in the form of a solid until it is sheared with sufficient stress to break down that structure. The ink then flows with a viscosity which decreases with increasing rate of shear.

The inks used for printing and other forms of reproduction (like xerography) are as varied and sophisticated in their composition as the paints discussed in §10.2.2. Indeed, some of the demands placed on the inks used in high-speed operations are more stringent than for any paint. The traditional printing inks and permanent black writing inks were made by making a complex of iron with some organic acid of somewhat uncertain composition. (Gallic or tannic acid were the usual materials and neither of them has a well defined formula.) Additional black pigments were added and the mixture of colloidal materials was suspended in a vegetable or mineral oil, using appropriate surface-active agents as stabilizers. They worked much like the oil-based paints: the oil reacted with oxygen from the air to form a polymer which bound the pigment to the paper. More modern inks use a variety of pigments, even to make basic black ink, and use polymeric materials in place of the oils as the binding agent in printing inks. The surface properties and the flow behaviour of liquid printing inks can be understood using the same principles as those for paints. The interaction between the ink and the paper is, however, much more significant in this case as the dynamics of that interaction extend from milliseconds to several minutes.

Printing inks dry by two mechanisms: evaporation of the solvent (or **vehicle** as it is called) and by the movement of the vehicle into the pores of the paper, leaving the pigment behind on the surfaces of the larger pores, as on a filter paper. Fig. 10.4 shows a simplified model of the process in a typical letterpress printing operation (Lyne and Aspler 1982). The plate or roller, with its load of ink applies a high pressure to the paper and effectively injects the ink into the pores. Wetting is not important in that process, and the advancing meniscus is concave. But as the pressure is removed and the film begins to split (1), an even image will only form if the ink wets the paper. The paper is in contact with the roller for a time of order a millisecond and the ink penetrates to a depth of the order of a micrometre. As the pressure wave moves away, the paper (which is also viscoelastic under these conditions) relaxes and draws the liquid (vehicle) into its pores. Note the reversal of the curvature of the menisci in (2). Only then do the finer capillaries come into play, in the seconds after the roller has passed, drawing the vehicle further into the pores to dry out the image (3). That process continues for

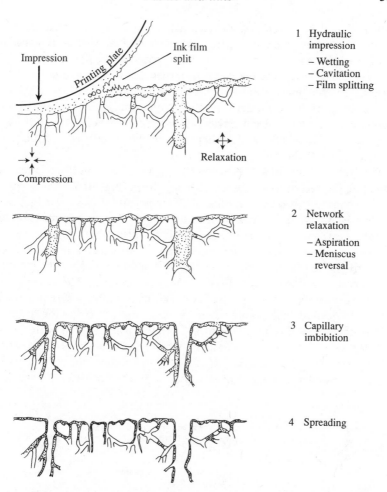

Fig. 10.4. A simplified model of ink impression and absorption. (1) The ink is hydraulically pressed into the pores. As the roller pulls away, the ink film splits by cavitation and rupture and the surface is thoroughly wet. (2) The paper relaxes and draws ink into the larger pores. (3) The capillary forces take over and draw liquid into the smaller pores, leaving most of the pigment behind. (4) The ink continues to spread slowly over all the surfaces. (From Lyne and Aspler (1982, p. 387) with permission.)

minutes after the roller has passed on. Finally, there is a slow spreading process, which can continue for days or even weeks after the printing process is complete (4). The crucial period occurs immediately after the roller leaves the paper, since it is then that the image is set, as the removal of the liquid increases the concentration of the pigment and hence its viscosity.

In high-speed printing operations, especially when less absorbent high-quality magazine paper is being used, the drying process is assisted by heating which may simply force the rapid evaporation of the liquid vehicle. In some cases, where the ink contains a thermosetting polymer, the heating actuates the polymerization to produce a solid ink.

The inks used in modern photocopying machines are of a special character. The more recent devices are 'plain paper copiers' and, as their name implies, they use standard paper; they also usually use a solid powder (called **toner**) as the ink.

The copy is made in a photocopier by projecting an image of the printed page (the **master**) onto a photoconducting drum (Fig. 10.5). The drum must first be charged up with a coating of electrons and this is done in a corona discharge. As the image strikes the drum, the light areas become conducting and the electrons at those points leak from the drum surface. The drum then revolves away from the master and comes into contact with a roller which contains the developer (ink or toner). It is actually a dry, black colloidal powder with a particle radius of a few micrometres. The charge on the powder is positive so it sticks to the drum only at points where the electrons remain. Those are the points where the light did not strike the drum.

The drum now has an image of the original page written with the black powder which is sticking to the drum wherever there was no light reflected from the master. It rotates a little further and comes into contact with the paper. The paper is charged negatively and so the positive ink particles are

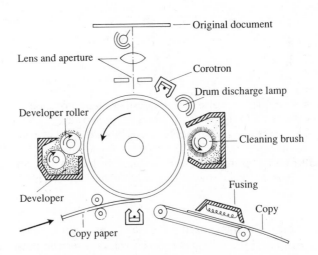

Fig. 10.5. Schematic diagram of a typical xerographic copier. Processing clements are clustered around the drum. (Reproduced with permission from Weigl (1982). Copyright American Chemical Society.)

transferred to the paper. The paper is then passed near a very hot roller which melts the ink into the paper surface. Meanwhile the drum continues its rotation, is cleaned of any remaining powder and discharged completely before being recharged by injection of ions from the corona wire. It is then ready to receive the next image.

The ink, or toner, is a key element in this process and its properties must be well controlled. It is charged during its manufacture with a sign opposite to that on the drum. The toner consists of irregular shaped particles of about 5–6 μm radius. The particles are made of a polymer resin in which is dispersed the pigment (colloidal carbon black for the usual black ink). The resin is a brittle material and the manufacture of the toner is done by 'jet-impaction' which breaks the resin into particles of the right size and also charges them up. The resin is also **thermoplastic** so it tends to flow when heated; it is this property which binds the pigment particles to the paper.

10.2.4 *Inking, painting, and coating by electrophoresis*

Early versions of these photocopying machines used a special paper and a liquid ink. To get the ink to dry quickly enough it was necessary to expose it to ultraviolet radiation and to incorporate special photochemical initiators to speed the drying process. There are also devices in which an image is first formed as a pattern of charges on a plate or paper. The latent image is then developed by transferring pigment particles to the surface from a colloidal dispersion in a liquid. The principle of the method (called **liquid immersion development** (Lyne and Aspler 1982) of LID) is shown in Fig. 10.6; it is, in effect, an electrophoretic development process (§8.4.1). The method can be

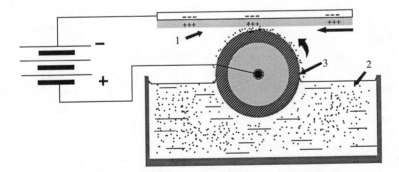

Fig. 10.6. Electrophoretic liquid development apparatus. The charge which carries the image is on the photoconductive insulator (1). The roller (3) picks up the pigment particles (2) and brings them near to the surface of the photoconductive insulator where the electric field transfers them to the surface. (Redrawn from Weigl (1982).)

312 *Applications of colloid science*

made to produce extremely high resolution, because it uses individual colloidal-sized particles to develop the image.

A typical developer is made up of a carbon pigment coated with a monolayer of a **'charge control agent'** dispersed in a hydrocarbon liquid. The charge control agent is usually a long-chain aliphatic salt (that is, a soap (§1.4.2)). The soap can undergo partial ionization in the hydrocarbon liquid. In this case it is the long-chain negative ions which go into solution to form the diffuse double layer, leaving behind the metal ion on the surface of the carbon black particle, which thus becomes positively charged. Only about 1 in 10 000 of the surface monolayer of soap molecules needs to ionise in order to produce enough charges (35–150) to allow a 0.5 μm particle to take part in the formation of the image. The electrophoretic mobilities (Lyne and Aspler 1982) in these systems are commonly of the order of 0.3 to 0.8 × 10^{-5} cm^2 V^{-1} s^{-1} and with an electric field of around 10^5 V cm^{-1} and electrode spacing of 1 mm, the development time is 50 ms or less. An analogous process is used on a much larger scale to coat and decorate motor car bodies. Layers of polymer latex, pigment, and corrosion protection materials, can be transferred from a colloidal dispersion to the surface of a metal plate by making it the electrode in a cell and using the process of electrophoresis (§8.4.1). Once the first layer of particles has been deposited, the next layer of the same material will encounter a repulsive force as particles of the same sign approach the surface. The applied voltage must be such that it is able to overcome that repulsion. The advantage of the method is that the developing film tends to be very uniform because the electric field is strongest where the film is thin so new particles tend to be pulled towards the thin parts of the film. The method can also get the coating into the tight corners and recessed areas which might well be missed by a spraying operation. It is also much safer and more economical to be applying the paint and other materials from solution where there is no loss of material into the atmosphere. The deposition can be done from an aqueous or nonaqueous solution.

A similar electrophoretic deposition process is used to produce the coating of phosphor on a television screen and the activated cathode in a cathode ray tube. It is also used to make thin rubber products like surgical gloves and condoms by the deposition of a rubber latex on a metal former of the desired shape.

10.3 Flotation of mineral ores

The most widespread and effective method of separation of a mineral from one of its ores is by **flotation**. Every year about a billion tonnes of ore is so separated and the value of the product is many billions of dollars.

A mineral ore usually consists of particles of some more or less pure mineral (like lead sulphide or manganese oxide) embedded in a matrix of some other, less valuable, material (often silica or a silicate). The size of the mineral particles might range from less than 1 μm upwards, but is usually from several millimetres up. The first step in the recovery process is to crush the ore to such a size (called the **release size**) that the individual particles are either mineral or worthless matrix but not both. The next step is to separate the mineral particles from the matrix material (called the gangue — pronounced 'gang') and that is where the flotation process is used.

The basic principle of flotation was probably discovered by gold miners. When sluicing or panning for gold one scoops up some river sand and silt into a shallow dish and then swirls it around and tries to pour off the solid material, leaving the heaviest (gold) particles to settle to the bottom of the pan where they should show up when most of the sand and silt has been decanted away. It turns out that if, in the excitement of the chase, one touches a gold particle, it will probably be lost with the lighter sand particles. Why is that?

The flotation process depends upon the attachment of an air bubble to a mineral grain. The combined density of the bubble and the grain is then usually low enough so that the pair can float to the surface of the liquid. In a mineral flotation cell, the air is bubbled into the base of the cell in which the mineral grains and the gangue are being kept in suspension by vigorous stirring. As the bubbles travel upwards they intercept the mineral grains and carry them up in the froth or foam. In the gold-panning operation, the air is pulled in by the agitation. Normally that doesn't matter because the gold particles will not become attached to the bubbles. They do so only when they are touched, presumably because they then pick up a film of oil from the fingers and this makes the surface hydrophobic.

When one wants to float a mineral deliberately, one may rely on its natural flotability if, like lead sulphide, it happens to be very hydrophobic. If, like many of the oxides, the mineral is naturally hydrophilic, then it becomes necessary to make the surface hydrophobic by adsorbing a surfactant onto the surface. One must choose conditions so that the surface of the worthless gangue material remains hydrophilic and that is where our knowledge of surface chemistry becomes very important.

A surface-active substance, added to a flotation cell to make the mineral float, is called a **collector**. For the sulphide minerals, if a collector is needed the usual choice is ethyl xanthate, which is made by reacting sodium ethanolate with carbon disulphide:

$$C_2H_5\text{-}ONa + CS_2 \rightarrow C_2H_5\text{-}O\text{-}C{=}S \text{ (sodium}$$
$$| \quad \text{ ethyl}$$
$$SNa \text{ xanthate).}$$

Though it has been in use for well over half a century, there is still some

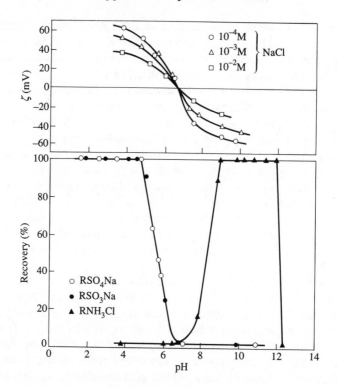

Fig. 10.7. The dependence of the flotation properties of goethite (FeO(OH)) on surface charge. Lower curves show the flotation recovery in 10^{-3} M solutions of dodecylammonium chloride, sodium dodecylsulphate and sodium dodecylsulphonate. (After Fuerstenau (1957).)

argument about the way this substance works. It is rather labile and readily oxidizes or decomposes into related materials, some of which may be involved in the flotation process.

The behaviour of oxide flotation agents is more easily understood, thanks largely to the work of Fuerstenau and his collaborators. Figure 10.7 shows the relation between surface properties and flotation recovery for the mineral goethite using a long-chain anionic or a long-chain cationic surfactant molecule as the collector. Notice (from the upper figure) that the surface charge on the goethite can easily be controlled using the pH. It is positive for pH values below about 7 and negative for higher values. Notice also that the collector which works best has the opposite sign of charge to the surface: the sulphate and the sulphonate work well on the positive surface and the substituted ammonium ion works best on the negative surface. The collector is absorbed with its headgroup attached to the mineral surface and the chain

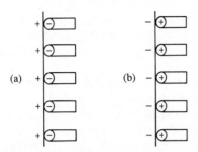

Fig. 10.8. Schematic picture of the adsorption of surfactants on goethite (FeO(OH)). (a) Long-chain sulphate at pH < 7. (b) Long-chain ammonium salt at pH > 7.

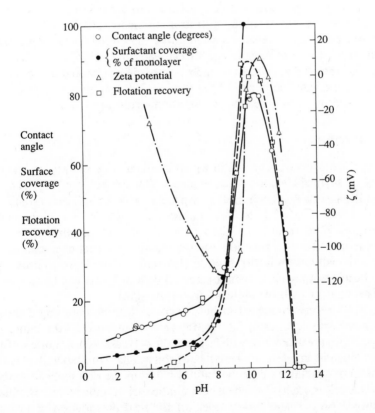

Fig. 10.9. Correlation of adsorption density, contact angles, and zeta potentials with flotation recovery of quartz with 4×10^{-5} M dodecylammonium acetate additions. (From Fuerstenau (1957).)

pointing into the solution (Fig. 10.8). That is why the surface of the particle becomes hydrophobic. The adsorption can occur only if there is an electrostatic attraction between the headgroup of the surfactant and the surface of the mineral grain.

The interaction requires that the film of liquid between the approaching bubble and the particle must be able to thin down and rupture so that the bubble is actually attached to the particle with no intervening layer of water. The details of the interaction process need not concern us at this point, but we can note that the surface charge effects, the adsorption coverage, and the contact angle (or wettability) of the grain all become optimal in the region where the particle becomes easily floated (Fig. 10.9).

The bubbles begin at the base of the flotation cell with a radius of a few millimetres, but by the time they have reached the surface they may be several centimetres in radius and are normally totally covered with the mineral. The foam is allowed to overflow into a receiver from which the mineral may be separated by filtration or centrifugation. The surface chemistry becomes more subtle where more than one mineral is present in the mixture after crushing. In the silver, lead, zinc mines in Broken Hill, Australia, the zinc and lead sulphides are readily separated using the difference in the hydrophobicity of the surface. Lead sulphide is able to float without assistance but the zinc requires some collector.

10.4 Detergency

One of the most obvious areas in which the science of surfaces is involved is that of surface cleaning or detergency. The subject stretches from the preparation of surfaces in industry for processes like painting and soldering, to the many aspects of domestic cleaning. The methods depend on the sort of surfaces involved and the likely form of the 'soil' or 'dirt'. (Detergent manufacturers tend to use the term 'soil' as an upmarket expression when what they really mean is dirt. Soil, as that term is used in agriculture, normally becomes dirty only when it is mixed with oil. That can happen when it makes contact with the natural oils on the skin).

Whether we are interested in hard-surface cleaning (washing dishes or floors) or cleaning fibres or fabrics, the general approach is the same. The 'dirt' particles are usually very different in the two cases. On a fabric surface, the dirt is usually made up of a mixture of soil (mineral matter) and oil (either a mineral or vegetable oil or the natural oil from the skin (called **sebum**)). On dishes it is more likely to be a combination of organic materials, chiefly denatured† protein and the products of the heat degradation of fats and

---

† A protein is said to be 'denatured' when it irreversibly loses its normal three-dimensional shape, usually as a result of some chemical reaction or heating or an adsorption process which disturbs the usual hydrogen-bonding structure.

sugars. The method of removing the dirt from the underlying surface is illustrated in Fig. 10.10.

The considerations involved in washing clothes will serve to illustrate the function of the various components of a typical clothes washing powder or liquid. The most important ingredient is, of course, the detergent. The dirt is almost always hydrophobic since if it is highly hydrophilic it will fairly readily disperse in water without the aid of a detergent. The function of the detergent is to adsorb onto all of the available surfaces and to penetrate even into regions where the only interface is between the dirt particle and the underlying fibre. The agitation of the wash then produces enough shearing action to tear the dirt particles from the fabric and, as they come away, more detergent is adsorbed on the freshly exposed surfaces. (Here the presence of the micelles acts as a reservoir for more detergent.) The detergent adsorbs with the hydrocarbon chain buried in the oily hydrophobic surface and the headgroup exposed. It is the repulsion between the head groups on adjoining surfaces which encourages the dirt particle to lift away from the fibre surface. The detergent also stabilizes the dispersion of oil and mineral particles, keeping them suspended in the water until the wash water is drained and the fabric is rinsed.

A washing powder may also contain some or all of the following:

1. A filler and drying agent like sodium sulphate which is slightly efflorescent and which helps to keep the powder running freely, both in manufacture and in use.

2. Some agent to remove calcium from hard water† (usually sodium hexametaphosphate $Na_6(PO_3)_6$ or a similar substance).

3. An anti-redeposition agent like carboxymethycellulose, a hydrophilic polymer, which covers the surfaces of both the fabric and the dirt particles and provides steric stabilisation (§2.4.1) in addition to the electrostatic stabilisation of the detergent.

4. A mild bleaching agent like peroxide or percarbonate: percarbonate decomposes to carbonate and gets rid of some of the calcium.

5. A brightening agent: a substance which absorbs in the near-ultraviolet and reradiates in the blue end of the spectrum can make an otherwise yellowish looking fabric look more white (in fact it can even be 'whiter than white' in the sense that it is reflecting more white light than is incident on it.

6. Perfumes and other additives for special purposes, like titania (TiO_2). TiO_2 is incorporated into synthetic fibres to make them white instead of

---

† Although modern detergents do not have insoluble calcium salts (like the old soaps) they still function less efficiently in the presence of the calcium ion.

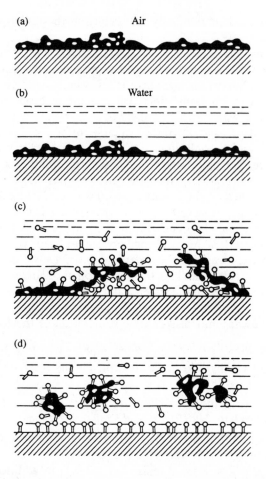

Fig. 10.10. Mechanism of action of a detergent in a fabric washing process. (a) The soiled surface. (b) Addition of water containing a surfactant. (c) The detergent becomes adsorbed onto all exposed surfaces. (d) The agitating action of the machine shears off some of the dirt and suspends it in colloidal dispersion. Note that the size of the surfactant molecules is grossly exaggerated here. The droplets of the oily material are huge compared with the size of the molecules. After this stage, more detergent penetrates into the interfacial region between the fibre surface and any remaining dirt particles. Another stabilizer (like carboxymethylcellulose) is often added to the mixture and it becomes adsorbed onto all the exposed surfaces, preventing the dirt particles from redepositing.

translucent or transparent. The washing process may remove some titania so there is an advantage in trying to replace some of it.

Some washing powders also try to overcome the problem of 'harshness' in the fabric by incorporating a **fabric softener**. This is another surfactant, usually cationic, which is more often incorporated in the rinsing cycle. It is necessary because modern detergents tend to be rather too efficient. They do not leave behind an oily film, like the old-fashioned soaps used to do, and as a result the fibres of the fabric have no lubrication. When the cloth article is crushed it feels harsh. Coating the surface of the fibres with a cationic surfactant provides that lubrication because the positive headgroups go in head down onto the negative surface and leave their hydrocarbon chains on the outside to provide an oil film. Soaps leave a similar lubricating film but also carry with them some precipitated calcium soap which builds up a loading on the fabric surface, indistinguishable from dirt.

10.5 Food colloids

Almost all of our foods are complex colloidal systems but their surface and colloidal aspects are often masked by the fact that the colloidal components are sealed inside cell walls. Obvious exceptions are milk and the various dairy products like custards, creams, and ice-cream. Jellies, junket, and all the various sauces, both sweet and spicy made from cream or milk, or flour are colloidal dispersions or gels.

A recent excellent introduction to food colloids (Dickinson 1992) describes the variety of colloidal structures in food thus:

The particles may be exactly spherical (gas bubbles, oil or water droplets) or approximately spherical (fat globules, protein or starch granules). They may occur in a wide range of sizes—from nanometres (surfactant micelles) to micrometres (emulsion droplets) to millimetres (foam bubbles). Particles may exist as isolated dispersed entities or they may be stuck together to form aggregates of various shapes, sizes and structures. Non-spherical particles may be needle or plate-like (ice, fat, or sugar crystals), fibrous or sheet-like (filaments, membranes) or deformed spheres (droplets, bubbles) distorted by mechanical forces or high density packing. Food macromolecules may be compact and highly organized (globular ptoteins) or chain-like and disorganized (many polysaccharides and denatured proteins). Molecular weights vary between tens of thousands of daltons (many proteins) to several millions of daltons (many polysaccharides and denatured proteins). Aggregates of macromolecules or particles may be so large as to span across macroscopic dimensions, thereby producing a gel-like network.

One of the most striking features of such foodstuffs is their rheological properties, which are significant because they often have a strong effect on

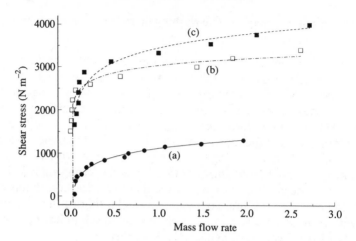

Fig. 10.11. Flow curves for butter (b) compared with (a) table margarine (high in unsaturates), and (c) cooking margarine (low in unsaturates). The mass flow rate measures the function $4Q/\pi a^3$ with Q in mg min^{-1} and a in cm. (See Exercise (4.6.5).) (Data from Sherman (1970).)

the taste and texture of our food. Mechanical properties, like the spreadability of butter or the toughness of meat, are quantified by using the techniques of rheology to which we referred in Chapter 4, but there are special problems in dealing with foodstuffs.

The behaviour of fats like butter and margarine, for example, depends strongly on the previous history of the sample, particularly the way it has been **worked** prior to the test. Working the fat, by, for example, extruding it repeatedly through holes in a plate, will reduce the hardness because it breaks up some of the crystal structure. Both margarine and butter consist of a dispersion of water in a continuous oil phase. In butter the oil is the natural fat mixture from milk, stabilized by milk proteins. In margarine, the oil is a combination of vegetable (and possibly animal) oils stabilized by synthetic or natural surfactants.

The spreading characteristics of butter, compared with those of the polyunsaturated vegetable oils like those used in modern table margarine, can be understood to some extent in terms of the softening temperature of the oil. Polyunsaturates have a lower softening temperature because the presence of the double bonds restricts the free rotation of the chain. The molecules cannot pack as well as the straight saturated chain and so the van der Waals interaction between molecules is not so strong. Figure 10.11 shows the flow curves for butter and table margarine.

Other dairy products like cream, ice-cream, and yoghurt have properties which are determined, at least in part, by the presence of a structure derived

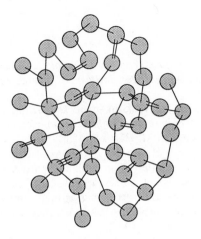

Fig. 10.12.   Subunit model of the casein micellar aggregate. The lines connecting the individual micelles represent bridges of colloidal calcium phosphate. Unconnected surface regions of micelles are rich in κ-casein, a form of the protein which does not interact with calcium. (After Dickinson (1992).)

from a number of related proteins referred to collectively as **casein**. Casein occurs in the form of micellar aggregates; Fig. 10.12 shows a schematic diagram of the arrangement in which it is believed to occur. The individual casein micelles are linked together by bridges of colloidal calcium phosphate with the aqueous medium able to enter the region between the micelles.

Ice-cream is simultaneously both an emulsion and a partly solidified foam. The incorporation of air into the emulsion of fat in water is essential to give it the smooth, creamy consistency; without the air it would also be too cold to eat comfortably. The bubbles of air are about 100 μm in size and the solid foam is held together partly by emulsified fat globules and partly by a network of small (about 50 μm) ice crystals dispersed in a sweetened aqueous macromolecular solution (Dickinson 1992). The conditions during formation and storage of the ice-cream prevent the sugars (sucrose and lactose) from crystallizing. The cooling is so rapid and the viscosity of the aqueous phase is so high that the sugars form a glassy solid. Thus ice-cream is 'not only a foam, but also an emulsion, a dispersion, and a gel; partially gaseous, partially liquid, partially crystalline, and partially glass-like' (Dickinson 1992).

For more details on the principles of colloid science as they apply to food systems, see Dickinson (1992).

References

Dickinson, E. (1992). *An introduction to food colloids*. Oxford University Press.

Fuerstenau, D. W. (1957). *American Institute of Mining Engineers Transactions*, **208**, 1365.

Lyne, M. B. and Aspler, J. S. (1982). Ink–paper Interactions in Printing — a Review. In *Colloids and surfaces in reprographic technology*, American Chemical Society Symposium Series 200, (ed. M. Hair and M. D. Croucher, Ch. 20, pp. 385–420.

Sherman, P. (1970). *Industrial rheology* p. 195. Academic, New York.

Underwood, S. and Jones, A. (1991). Some colloidal aspects of the rheology of paint. *Chemistry in Australia*, January–February, 29–31.

Weigl, J. W. (1982). Interfaces in electrophotography. In *Colloids and surfaces in reprographic technology*, American Chemical Society Symposium Series 200, (ed. M. Hair and M. D. Croucher), Ch. 8.

Answers to Exercises

Chapter 1

1.1.2   Total area $= 6 \times 10^4$ cm^2; 6×10^5 cm^2; 6×10^6 cm^2
        Surface energy per particle $= 4.2 \times 10^{-13}$ J; 4.2×10^{-15} J; 4.2×10^{-17} J
        All much larger than kT. Total surface energies for all the particles are 0.42 J, 4.2 J, and 42 J respectively.

1.1.3   Area per gram $= 7.1$ m^2/g.
        Volume fraction $= 0.030$.

1.1.4   Area per gram $= R(1 + 2/R)/[\rho a]$

1.4.4   0.093 mmoles.

1.4.5   The radical is about 100 times more likely to collide with a micelle than it is with a droplet.

1.4.6   The radical is only about 5 times more likely to collide with a molecule as it is with a droplet. Collision with the micelle is the most likely.

1.4.9   3.1 millimoles/100 g. (Usually written as 3.1 meq/100 g.) (One equivalent is one mole of univalent ion in this context so 1 meq is the same as 1 millimole.)
        Area per charge $= 0.77$ nm^2. Charges are about 0.9 nm apart.

1.4.10 (a) 1200 negative charges per particle.
            6240 positive charges per particle.
            Net charge is 5040 positive charges per particle.
        Similarly for the other pHs:

pH	6	8	10
− ve charges	11000	12000	12000
+ ve charges	6250	3120	620
Net charge	− 4750	− 8880	− 11380
μC/cm^2	− 0.60	− 1.13	− 1.45

pH at the iep is approximately 5.

1.5.1   Total area $= 231$ cm^2/L.
        A 1 L flask would have an area of the order of 6×10^2 cm$^2 = 600$ cm^2.
        To cover 115 cm^2 of surface requires 5.6 μg.

Chapter 2

2.1.1 $k = 1.107 \times 10^{-23} \, \text{J K}^{-1}$. $N_A = 7.51 \times 10^{23}$.
          For $r = 22$ nm, $k = 1.27 \times 10^{-23}$ and $N_A = 6.53 \times 10^{23}$.
2.1.2 $N_A = 6.92 \times 10^{11}/D$ for D in SI units. Values between 4.74 and 11.4×10^{23}.
2.1.3 $D = 2.44 \times 10^{-12} \, \text{m}^2 \, \text{s}^{-1}$. Apparent (diffusion) radius of chloride ion $=$
          0.122 nm.

Chapter 3

3.1.1 $\bar{d} = 307.4$ nm $\sigma = 103$ nm; $\sigma^2 = 1.06 \times 10^4 \, \text{nm}^2$. Mode ≈ 275 nm; difference $\approx$
          32 nm.
3.1.3   (a) $K = (\pi/6)^{1/3}$ (b) $K = (\pi \rho/6)^{1/3}$
3.1.4 $\bar{d} = 307.4$ nm $\bar{d}_{NA} = 324.2$ nm.
3.1.5 F_{max} occurs when $\bar{d}_i = \bar{d}$ and is $3N/4\bar{d}$.
3.1.6 $\bar{d}_{NV} = 340$ nm. (If the $(\pi/6)^{1/3}$ term is included then $\bar{d}_{NV} = 274$ nm.)
3.1.7 $[7\sqrt{5}/25]$ i.e. 62.6% of the sample lies in this region compared with 68.2%
          for the normal distribution.
3.2.1 $\lambda = 1.23 \times 10^{-11} \, \text{m} \approx 0.012$ nm. This is obviously not the factor which limits
          the resolution of the electron microscope. Rather it is distortion which occurs
          in the magnetic and electrostatic focusing devices.
3.3.1   For $r = 1 \, \mu\text{m}$, $t = 3.1 \, \mu\text{s}$; for $r = 0.1\mu\text{m}$, $t = 31$ ns; for $r = 0.01 \, \mu\text{m}$, $t = 0.31$ ns.
3.3.2 $t = 14.158$ hr $= 14$ hr 9.5 min.
3.6.3 $\Delta P = 4 \times 10^7 \, \text{Pa}$ ($= 400$ atm).

Chapter 4

4.2.4   Wall shear rate $= 509 \, \text{s}^{-1}$.
          (Values from $500\text{–}2000 \, \text{s}^{-1}$ are common for wall shear rates in capillary
          viscometers at normal flow rates. It should be noted that such shear rates
          may be sufficient to tear a long chain polymer molecule apart. Forcing high
          molecular weight polymers through capillaries can cause considerable
          degradation.)
          The shear stress at the wall $= 0.581 \, \text{N m}^{-2}$ (or Pa). Note how small this is
          compared with atmospheric pressure ($\approx 10^5$ Pa). A liquid cannot support
          any shear stress at all without undergoing flow.
4.3.1   3.33.
4.3.3   All of the shear data can be plotted reasonably well on a linear log–log plot.
          The regression line is not a bad fit but it is clear that the data follow a curve.
          The data can be broken into low, medium and high shear regions and these
          show better fits to eqn (4.24) K and n parameters can be obtained by linear
          regression analysis for these plots and the values obtained are shown in the
          table below.

	K	n
All data	0.7405	0.859
Low shear rate	0.7761	0.983
Intermediate shear	0.825	0.867
High shear rate	1.312	0.699

To check the Ellis model we need values of η as a function of shear rate S and these are obtained from $\eta = S/\dot{\gamma}$ from which we obtain η_0. The data are rather scattered at the low shear end but a value of 0.8 Pa s appears to be appropriate. It is apparent that the data have not been taken to high enough shear rates to reach η_∞. That parameter would have to be less than 0.25 Pa s and would not affect the data in the figure. As can be seen from Fig. 4.12 in the text, the Ellis and Meter models are identical until the viscosity begins to level off at the high shear rate end.

4.6.6   For the non-Newtonian fluid, the ratio $[p_{high}/p_{low}]^{1/0.3} = 4$ and so $p_{high}/p_{low} = 4^{0.3} = 1.52$. The increase in pressure drop required to produce a four-fold increase in the flow rate is much smaller for this fluid.

Chapter 5

5.2.2   (i) $1.4 \times 10^6 \, \text{N m}^{-2} \simeq 14 \, \text{atm.}$;   (ii) $\simeq 140 \, \text{atm.}$

5.4.1   The lowering is about 0.001% and $p''/p^0 = 0.9999898$.
        The capillary rise $h = 143 \, \text{mm}$.

5.6.4 $r_c = 0.709 \, \text{nm}$. Volume $= 1.5 \times 10^{-27} \, \text{m}^3$.
        The volume of a water molecule $\approx 3 \times 10^{-29} \, \text{m}^3$ so this sphere would contain about 50 molecules.

Chapter 6

6.1.1 $\Delta H_{ads} = -905R = -7.52 \, \text{kJ/mol.}$; $45 \, \text{m}^2 \, \text{g}^{-1}$.

6.1.2   (b) $V_{mon} \approx 25 \, \text{cm}^3$ from linear plot. V_{mon} is $27.9 \, \text{cm}^3$ from eqn 6.11.
        Area $= 120 \, \text{m}^2/\text{g}$

6.1.3   Area $= 33 \, \text{m}^2 \, \text{g}^{-1}$ for A and $82 \, \text{m}^2 \, \text{g}^{-1}$ for B.

Chapter 7

7.2.2   (i) $0.01 \, \text{M Na}_2\text{SO}_4$: $I = 0.03 \, \text{mol L}^{-1}$; $\varkappa^{-1} = 1.76 \, \text{nm}$.
        (ii) $0.015 \, \text{LaCl}_3$: $I = 0.210 \, \text{mol L}^{-1}$; $\varkappa^{-1} = 0.664 \, \text{nm}$.
        (iii) $3 \times 10^{-3} \, \text{M Ca(NO}_3)_2$: $I = 9 \times 10^{-3} \, \text{M}$; $\varkappa^{-1} = 3.205 \, \text{nm}$.
        (iv) $0.025 \, \text{M Fe(NH}_4)(\text{SO}_4)_2$: $I = 0.225 \, \text{M}$; $\varkappa^{-1} = 0.641 \, \text{nm}$.
        (v) $10^{-4} \, \text{M La}_2(\text{SO}_4)_3 + 5 \times 10^{-4} \, \text{M NaNO}_3$: $I = 2 \times 10^{-3} \, \text{M}$; $\varkappa^{-1} = 6.8 \, \text{nm}$.

7.3.1 $\sigma_0 = 0.539\ \mu C\ cm^{-2}$

Number of silver ions per $cm^2 = 3.36 \times 10^{12}$.

The number on the crystal surface if they are on a square lattice of side 0.4 nm is 6.25×10^{14} ions/cm^2.

Number added to generate a surface potential of 60 mV is only 0.5% of the number already present. This is why the activity of the silver ions on the surface can be assumed to remain constant.

Chapter 8

8.1.1 Surface charge density (cation exchange capacity) = 3.72 mmole/100g = 20.5 $\mu C/cm^2$. [This corresponds to an area per charge of 0.78 nm^2, which is a very high value. Few, if any, systems are able to pack ions in at a surface density higher than one per 0.5 nm^2.]

8.4.6

a (μm)	0.2	0.8	1.2	5
X	0.950	0.4637	0.3122	0.07289
Y	0.124	0.2853	0.2347	0.06978
$\|G\|$	0.958	0.5444	0.3906	0.1009

Thus at 1 MHz frequency this inertia effect reduces the mobility from its d.c. value by about 4% for a 0.2 μm particle and by 90% for a 5 μm particle.

8.6.1 In 0.001 M KNO_3 $\varkappa$ is 0.1040 nm^{-1}. Then $a = 9.62$ nm. An estimated value of around 1.05 is satisfactory for $f(\varkappa a)$. [The correct value at this value of $\varkappa a$ is 1.027.]

8.6.2 At 10^{-2} M: pH = 8.5; $\zeta = -20$ mV. $\sigma_d = -0.47\ \mu C\ cm^{-2}$.

At pH 10 $\zeta = -35$ mV, $\sigma_d = -0.863\ \mu C\ cm^{-2}$.

This is about 10% of the quoted surface charge of $-10\ \mu C\ cm^{-2}$. [The figure of $-70\ \mu C\ cm^{-2}$ in earlier print runs is a misprint. Such extreme values do occur on the clay mineral kaolinite but only rarely in other systems. When such large values do occur they may signify a porous surface though not in the case of kaolinite.]

At 10^{-3} M, the largest ζ potential is only -50 mV corresponding to $\sigma_d = -0.42\ \mu C\ cm^{-2}$.

Chapter 9

9.2.2 $A_{131} \simeq (A_{11}^{1/2} - A_{33}^{1/2})^2$

$A_{11}=$	1	2	3	4	5	6	7
$A_{131}=$	1.983	0.988	0.457	0.01667	0.0297	0.0017	0.0564

$A_{11}=$	8	9	10	11	12	$\times 10^{-20}$
$A_{131}=$	0.1765	0.3501	0.568	0.825	1.115	

9.2.3 $f(3) = 0.59$ in both cases. $f'(p) = -0.14$ for $p \leqslant 3$ and -0.12 for $p \geqslant 3$. So there is a difference in slope of about 15% between the two functions.

9.3.3   For As_2S_3 $z=1$; $A = 4.07 \times 10^{-19}$ J. For $z=2$ $A = 4.31 \times 10^{-19}$ J. For $z=3$ $A = 3.66 \times 10^{-19}$ J.

For the Au sol $z=1$ $A = 6.19 \times 10^{-19}$ J; $z=2$ $A = 6 \times 10^{-19}$ J; $z=3$ $A = 1.4 \times 10^{-18}$ J.

For $Fe(OH)_2$ $z=1$ $n^0 = 10$ mM; $A = 9.4 \times 10^{-19}$ J. $z=2$ $n^0 = 0.21$ mM; $A = 8.1 \times 10^{-19}$ J.

9.3.6 $V_R = 2\pi \epsilon a \psi_d^2 \ln[1 + \exp(-\varkappa H)]$ (a) For $c = 10^{-4}$ M NaCl: $\varkappa = 0.03288$ nm^{-1}.
                                                     (b) For $c = 10^{-2}$ M; $\varkappa = 0.3288$ nm^{-1}.

9.3.7 V_A (in J) $= -2.5 \times 10^{-18}/H$ (nm)

Note that the 10^{-4} M system has a large positive repulsion energy at intermediate distances so that would be a stable sol. The 0.01 M system, on the other hand, has just dropped below the zero line, indicating that it has just entered the region of instability. Such a sol would be just entering the region of rapid coagulation where every collision would result in a doublet.

9.3.8   Depth of well $\approx -1.2 \times 10^{-19}$ J. The value of kT at 25°C is 4.12×10^{-21} J. The well is about $30\,kT$. Gravitational potential energy for a 0.5 μm particle $= 1.54 \times 10^{-14}\,h$ J. So $h \approx 7.8 \times 10^{-6}$ m $\approx 8\,\mu$m.

The attractive force is quite strong enough to dominate over gravity for structures which are several times larger than a single particle. (This is even more obvious if one takes into account the buoyancy of the water.) Some of the stress is also borne by the particles themselves as they undergo some deformation under the load.

9.4.1   Total thickness of water layer is 26.6 nm. No. of layers $= 5.47 \times 10^5$. The swollen block will then be 1.51 cm in thickness.

INDEX

Bold numbers refer to major sections, *italics* to subsections. The letter 'f' refers to a figure and 't' to a table. Consecutive page occurrences are not indexed unless a figure is involved.

phosphate, calcium 222, 321
phosphor 312
photoactivity of titania 304
photochemical effects 190, 311
photocopier operation 310
photon correlation spectroscopy *44, 88*
physisorption 165
piezoelectric crystal transducer 247, 284
pigments 54, 302–12
pipeline flow 119
plane, crystal 17
  adsorption 179
  slipping 233, 236
plant growth 4, 20, 52, 153
plastic flow 99, 107, *114*, 122, 126, 308
plasticizers, as impurity source 25
plastics 302
platelets, clay, separation of 283
plates, flat 6, 58, 107
  interaction between 264, 271, 274
platinum electrode 195f, 201, 244, 252
plug flow 238
plug, porous
  electro-osmosis in 235f
  streaming potential in *252*
PMMA 67, 229
point of zero charge, *see* p.z.c.
Poiseuille flow 90, 105, 242, 250
Poisson's equation 205, 236
Poisson–Boltzmann equation *204, 207,*
  266
polar groups, influence of 12
polarizability 42, 81, 270, 273, 289
polarization
  electrical 196, 213
    anodic 213
    cathodic 199
  of light 43f
pollen grains, Brownian motion 33
pollution 5
polyacrylic acid 116f
polyalcohols 304
polydisperse 59, 62, 96
polyethylene, use in replication 71f
polymer
  adsorption 25, 56
  latex 22, 229, 300, 312
    preparation of *20*
    rheology of 112f
polymerization, emulsion *20*
polymer–polymer interactions 54
polymer–solvent interactions 54
polymethylmethacrylate, *see* PMMA
polysaccharides 5, 223, 302, 319
polystyrene particles 71, 85, 92
polyunsaturated oils 12, 320
population, statistical 61, 65

pores, flow in 90
  in adsorption 141, 172
  membrane 27–31
  soil 153
porosimetry 153
porous layer 293
  plug 252
    electro-osmosis in 234, 247
    streaming potential of 252
  surface 302, 308
potential, electrical
  at flat surface **203**
  between plates 265f
  in double layer 209f
  influence on coagulation 267
  of diffuse double layer 213
    on oxides 222
  see also electrical double layer, zeta
    potential
potentiometer 196, 200
powder, adsorption on 174
  sintering of 149
  toner 310
precipitation, in colloid preparations 8, 10,
  49
pressure
  capillary 149, *151*
  excess
    inside a drop 138
    in small particles 149
  hydrodynamic 105
  hydrostatic 123
    in electrokinetics 249, 252
  inside bubble 135, 141, 146, 182
printing ink, rheology of 4, 302, 307–10
probability distribution 35
profile
  of bubbles and drops 183
  velocity, in capillary 89, 104, 125f, 237f,
    242f, 250f
prolate spheroid 58f
propagation time 272
protection 54
  against corrosion 282, 303, 312
protein 76, 80, 118, 223, 241, *254*, 300,
  316, 319
proton. activity 224
  adsorption on clay 52
  surface dissociation 228, 255
pseudoplasticity 107f, *115*, 118, 126, 130,
  300, 305
pulp, paper 115, 300
pulse counting 78
pulse, ultrasonic 247
purification
  of colloids **26**
  of water 28, 56